Nanotechnology in Medicine

Emerging Applications

NANOTECHNOLOGY IN MEDICINE

Emerging Applications

Gene Koprowski
LLM, MA, BA, Med. Dipl.

Nanotechnology in Medicine: Emerging Applications

First published by
Momentum Press ®, LLC
222 East 46th Street
New York, NY 10017
www.momentumpress.net

ISBN-13: 978-1-60650-248-8 (paperback, softcover)

ISBN-10: 1-60650-248-4 (paperback, softcover)

ISBN-13: 978-1-60650-250-1 (e-book)

ISBN-10: 1-60650-250-6 (e-book)

Cover design by Jonathan Pennell
Interior design by J. K. Eckert & Company, Inc.

10 9 8 7 6 5 4 3 2 1

Printed in the United States of America

For Nancy Bruening and Katherine B. Koprowski

Contents

PREFACE

How Nanotechnology Is Revolutionizing Medicine

The world is indeed getting smaller. The dimensions of nanotechnology are shrinking at a rather rapid rate. Consequently, more innovations are happening at the cellular, molecular, and even the atomic level. That's the definition of nanoscale: *the science of manipulating materials on an atomic or molecular scale especially to build microscopic devices.*[1] As scientific understanding grows, it is now possible to create the smallest devices and applications to help in a variety of medical fields.[2]

Nanotechnology is becoming vital to modern, allopathic medicine. Small nano devices that are being developed right now can enter the body and look around, and help with diagnostics, in ways that doctors could only dream of a decade ago.[2]

Nanotechnology is a modern scientific field that is constantly evolving as investor and academic interest continues to increase and as new research is developed. The field's simplest roots can be traced to 1959, but its primary development occurred more recently.[3] Specific scientific achievements, e.g., the invention of the scanning tunneling microscope (STM), show how the industry has been shaped by the the initial vision of molecular manufacturing as detailed by the physicist Richard Feynman in 1959.

> But I am not afraid to consider the final question as to whether, ultimately—in the great future—we can arrange the atoms the way we want; the very atoms, all the way down!
>
> —Richard Feynman
> *There's Plenty of Room at the Bottom*

As a physicist at CalTech, Feynman's talk, although it never included the word *nanotechnology,* suggested that it would be possible to precisely manipulate atoms and molecules. He also thought that it was possible to create *nanoscale* machines, through a cascade of billions of factories. According to Feynman, these nano factories would be smaller-scale models of machine hands and tools. These tiny "machine shops" would then eventually be able to create billions of smaller factories. Speculating further, he also opined that several factors uniquely affect developments on the nanoscale, i.e., as the scale got smaller, the force of gravity would become more negligible, while both Van Der Waals attraction and surface tension would become much more important. Feynman's talk has been viewed as the first academic lecture to dealt with the main tenet of nanotechnology: the direct manipulation of individual atoms, e.g., molecular manufacturing.

"The revolutionary Feynman vision launched the global nanotechnology race," according to physicist Eric Drexler.

Way before atomic force microscopes were developed, Dr. Feynman offered these wild ideas to his peers. He chose to deal with a "final question" that wasn't fully realized until the last decade. Ultimately, then, it was during these two decades that the term *nanotechnology* was devised, and researchers, starting with Eric Drexler, built up this field from the foundation that Feynman imagined in 1959. Some researchers, such as Chris Toumey, downplay the importance of Feynman's talk. Using evidence from its citation history, Toumey sees "There's Plenty of Room at the Bottom" as a "founding myth" that served to directly influence only Drexler but not others who also affected the development of nanotechnology. Whatever the case, though, ultimately, it is certain that Feynman directly influenced Drexler's own research, which thus indirectly influenced nanotechnology as a whole. *Supra.*

By 1979, Eric Drexler encountered Feynman's paper on atomic manipulation and "nano factories." The physicist's ideas inspired Drexler to put these concepts into action by further developing Feynman's vision of molecular manufacturing with current developments in understanding protein function. Drexler's primary objective was to build upon the physicist's foundation. The field of nanotechnology was created with the publication of Drexler's landmark work, *Engines of Creation: The Coming Era of Nanotechnology,* with a foreword by MIT's Marvin Minsky.[4]

Drexler discusses the development of molecular manufacturing as a "process" of fabricating objects with atomic specifications employed by specially designed protein molecules. Drexler suggests this will lead to the design of molecular machinery with the ability to position reactive groups with atomic precision. *Supra.* Therefore, he claims that molecular manufacturing and the construction of "nanomachines" are the products of an analogous relationship "between features of natural macromolecules and components of existing machines."

This book explores the implications—and developments—that stem from the pioneering theoretical works of Feynman and Drexler.

The following are dozens of ways that nanotechnology is revolutionizing medicine today, which are touched on in the following pages. *Supra.*

1. Nanobots: These robots could be used to perform a number of functions inside the body and out. They could even be programmed to build other nanobots.
2. Nanocomputers: To direct nanobots in their work, special computers will need to be built. Efforts to create nanocomputers, as well as the movement toward quantum computing, are likely to continue to provide new possibilities for medicine.
3. Cell repair: Cells are so incredibly small and difficult to repair. But nanotechnology could provide a way to get around this. Small nanobots or other devices could be used to manipulate molecules and atoms on an individual cell level.
4. Cancer treatment: The small, specialized functions of some nano devices could be directed precisely at cancer cells. Current technology damages the healthy cells surrounding cancer cells as well as destroying the undesirables. With

nanotechnology, it is possible that cancer cells could be targeted and destroyed with almost no damage to surrounding tissue.

5. Aging: Nano devices could be used to erase some of the signs of aging. Lasers can reduce the appearance of age lines, spots, and wrinkles. With nanotechnology, it is possible that these signs could be done away with completely.
6. Heart disease: Nanobots could perform a number of heart-related functions in the body. The repair of damaged heart tissue is only one possibility. Another option is to use nano devices to clean out arteries.
7. Implanting devices: Researchers say it might be possible to send a nanobot to build the structures inside the body.
8. Virtual reality: Physicians can explore the body more readily with the help of a nanobot injection, creating a virtual reality that would help medical professionals practice operations before they do them.
9. Gene therapy: Nanotechnology will be small enough to enter the body and even redesign the genome. This could alter a number of conditions and diseases. Nanobots could be qualified for swapping abnormal genes with normal genes and performing other functions.
10. Drug delivery: Systems that automate drug delivery can help increase the consistency associated with providing medication to those who need it when they need it. Drug delivery systems can be regulated using nanotechnology to

ensure that certain types of medications are released at the right place and time.

11. Nanotweezers: Devices designed to manipulate nanostructures can be used to move nano devices around in the body or position them prior to insertion. Nanotweezers are usually constructed using nanotubes.
12. Stem cells: Nanotechnology may actually help adult stem cells transform into the types of cells that are actually needed. Nanotubes already help adult stem cells turn into functioning neurons in brain damaged rodents.
13. Bone repair: Accelerating bone repair using nanotechnology is already happening. Nanoparticles made up of different chemical compositions can help knit bones back together and can even help with spinal cord injuries.
14. Imaging: Nanotechnology provides advancements in medical imaging by allowing a very specific and intimate peek into the body. These devices result in molecular imaging that can lead to better diagnoses of a variety of diseases.
15. Diabetes: Nanotechnology is providing a way for diabetics to use lenses to check their blood sugar. Nanotech contact lenses actually change color to indicate blood sugar level.
16. Surgery: Nanosurgery is possible using some lasers, as well as nano devices that can be programmed to perform some surgical functions. Performing surgery at the smallest level can have a number of benefits.
17. Seizures: Nanochips being developed can help control seizures. The chips analyze brain signals and then do what is

needed to adjust the brain so that epilepsy can be better controlled.

18. Sensory feedback: It is possible to use nanotechnology to increase sensory feedback. Nanochips provide the opportunity for electrical impulses to be intercepted and interpreted, replacing damaged nerve tissue function.
19. Limb control. Nanotechnology is helping to revolutionize the way paralysis is handled. There are some attempts to use nanochips that can help those who have lost limb control to use their minds to send signals to move limbs.
20. Medical monitoring: Small nanochips implanted in your body could monitor your health and body systems and then send feedback to your computer or other device.
21. Medical records: Nanotech can be used to send information to your health care providers and increase the efficiency of electronic medical records.
22. Disease prevention: With the correct programming, it should be possible to help patients avoid some diseases, repairing problems before they become serious. They may even be able to help prevent chronic problems.
23. Prenatal: Nanotechnology can help with prenatal diagnosis. Getting inside the uterus and the fetus without causing trauma can be beneficial to prenatal health, and nanotechnology can also help potentially repair problems in the womb.
24. Individualized medicine: Nanotech is making medicine more personal. Being able to accurately work up your

genome can help health providers more precisely pinpoint the proper treatments and tweak a treatment plan.

25. Research: Nanotech is advancing medical research, providing the tools that can help us learn more about the body and how it functions, as well as providing insight into chemistry and physics, which provide the building blocks for the genome.

References

[1] http:www.merriam-webster.com.

[1] *Future Medica: The future of medicine and biotechnology*, Jan. 19, 2010.

[2] Connexions: The Early History of Biotechnology, http://cnx.org.

[3] Drexler, K. Eric. 1987. *Engines of Creation: The Coming Era of Nanotechnology*. New York: Anchor Books, 320 pp.

INTRODUCTION

Nanotechnology in Medicine: Emerging Applications

Long a concern of futurists, nanotechnology is today transforming medicine. Nanotech concerns the manipulation of structures and properties at the nano scale, often at dimensions that are as thin as a fraction of a human hair. Nanotechnology is the basis for new, more effective drug delivery systems and is in early-stage development as scaffolding in nerve regeneration research. The National Cancer Institute has founded the Alliance for Nanotechnology in Cancer in the hope that investments in this branch of nanomedicine could lead to breakthroughs in terms of detecting, diagnosing, and treating various forms of cancer.

Nanotechnology medical developments are experiencing a wide variety of uses and could potentially save a great number of lives. Nanotechnology is not just used in passive structures but also in active structures, through more targeted drug therapies or "smart drugs." New drug therapies have already been demonstrated to cause fewer side effects and be more effective than traditional therapies. Nanotechnology is also assisting in the formation of molecular systems that are strikingly similar to living systems. These molecular structures are poised to be the basis for the regeneration or replacement of body parts that are currently lost to infection, accident, or disease. These developments are encouraging nanotechnology not only in terms of research and development but also in determining a means of oversight.

The number of products approaching the Food and Drug Administration (FDA) approval and review process is likely to grow as time moves forward and as new nanotechnology medical applications are developed.

Current medical applications include:

- Appetite control
- Cancer
- Cholesterol
- Drug development
- Imaging
- Medical tools
- Bone replacement
- Chemical substitutes
- Diagnostic tests
- Hormone therapy
- Immunosuppressant

Let's take a quick look at one technology in each of the areas cited above.

APPETITE CONTROL

Megace ES

Par Pharmaceutical Companies, Inc. (USA)

This drug is designed to stimulate appetite for "treatment of anorexia, cachexia, or an unexplained, significant weight loss in patients with a diagnosis of acquired immunodeficiency syndrome (AIDS)."

"Utilizes Elan's nanocrystal technology delivery system to improve the rate of dissolution and bioavailability of the original megestrol acetate oral suspension."

"Nanocrystal particles are small particles of drug substance, typically less than 1000 nanometers (nm) in diameter, which are produced by milling the drug substance using a proprietary, wet-milling technique. The nanocrystal particles of the drug are stabilized against agglomeration by surface adsorption of selected *generally regarded as safe (GRAS)* stabilizers. The result is an aqueous dispersion of the drug substance that behaves like a solution—a nanocrystal colloidal dispersion, which can be processed into finished dosage forms for all routes of administration."

✓ Approved by the FDA in July 2004.

Source: *http://www.elan.com/EDT/nanocrystal_technology.*

CANCER

Abraxane

American Pharmaceutical Partners, Inc. (USA)

This is an anticancer drug used to treat advanced breast cancer, described as an "albumin-bound form of paclitaxel with a mean particle size of approximately 130 nm."

✓ Approved by the FDA in January 2005.

Source: *http://www.fda.gov/cder/foi/label/2005/0216601bl.pdf.*

CHOLESTEROL

TriCor
Abbott Laboratories (USA)

This is a cholesterol-lowering drug that employs Elan's nanocrystal Technology to make it more easily administrable.

"Nanocrystal particles are small particles of drug substance, typically less than 1000 nm in diameter, which are produced by milling the drug substance using a proprietary, wet-milling technique. The nanocrystal particles of the drug are stabilized against agglomeration by surface adsorption of selected GRAS stabilizers. The result is an aqueous dispersion of the drug substance that behaves like a solution—a nanocrystal colloidal dispersion, which can be processed into finished dosage forms for all routes of administration."

✓ Launched in December 2004.

Source: *http://www.elan.com/EDT/nanocrystal_technology.*

DRUG DEVELOPMENT

Controlled Flow Cavitation
Five Star Technologies (USA)

"Technology controls the location, size, density, and intensity of implosion of bubbles in the cavitation zone to create optimum process conditions. This ability to harness the force of cavitation yields superior results when the controlled energy release is applied to nanomaterials synthesis and to the development of fine emulsions and dispersions for advanced applications."

It has been applied to a range of products from appetite control spray gels to metal oxide catalysts.

✓ First commercial platform introduced in 2005.

Source: *http://www.fivestartech.com/technology.*

IMAGING

TriLite Technology
Crystalplex Corporation (USA)

This technology is used to create nanocrystals and nanoclusters for imaging and diagnostics.

"TriLite™ alloyed nanocrystals are of relatively uniform size: approximately 6 nm… [A]lloyed nanocrystals can be produced in stable blue and blue green colors (less than 525 nm)… [C]rystals can also be made in the far-red region of the spectrum (greater than 660 nm) with little loss in stability.

"TriLite™ nanoclusters are aggregates of 8 to 12 individual nanocrystals. These nanoclusters are 40 to 50 nm in size and are functionalized on the surface with carboxyl groups using a proprietary Crystalplex technology."

The products are available in many sizes and colors and can also have nearly any biological probe bound to them.

✓ Available through company website.

Source: *http://www.crystalplex.com/.*

MEDICAL TOOLS

EnSeal Laparoscopic Vessel Fusion System

SurgRx, Inc. (USA)

"The SurgRx EnSeal Tissue Sealing and Hemostasis System allows surgeons to seal and transect small to large vessels, large pedicles and tissue bundles to achieve surgical hemostasis."

"With Smart Electrode Technology, EnSeal instruments adjust and dose energy simultaneously to various tissues, each with its own impedance characteristics. This proprietary electrode consists of millions of nanometer-sized conductive particles embedded in a temperature-sensitive material. Each particle acts like a discrete thermostatic switch to regulate the amount of current that passes into the tissue area with which it is in contact. EnSeal™ works equally well when sealing arteries and veins, and transecting fatty tissue, small ligaments, and connective tissue." ”

✓ Launched commercially in March 2004.

Source: *http://www.surgrx.com.*

BONE REPLACEMENT

Vitoss

Orthovita (USA)

"3-D ß-TCP scaffolds for use in repairing bone defects that are composed of highly porous, 100 nm-sized particles.

"Nano-sized particles enhance resorption and new bone growth."

✓ At least five Vitoss-containing products have been commercialized since February 2004.

Source: *http://www.orthovita.com/products/vitoss/overview.html.*

CHEMICAL SUBSTITUTES

Neowater

Do-Coop Technologies Ltd. (Israel)

"Broad enabling platform upon which research, diagnostics, biotech, and pharma companies can obtain disruptive price-performance results, yet with a graceful implementation."

"Neowater enables superior water-based biocatalysts, solvents, reagents, media enhancers, and buffers that maximize the efficiency of new and existing products and processes."

"Neowater™, with its stable system of largely hydrated nanoparticles, like nonionic detergent derived micelles, reduces the entropy of aqueous solutions. In addition, by design, it exhibits both hydrophilic and hydrophobic properties."

✓ Made generally available in February 2004.

Source: *http://www.docoop.com.*

DIAGNOSTICS

Microarrays

CombiMatrix Corporation (USA)

"Semiconductor-based array technology that enables the preparation of materials with nanoscale control... [and] allows for the parallel synthesis of large numbers of nano-structured materials. These materials can then be tested using the same chip-based technology."

The technology can be used for rapid analysis of samples, detection of disease, etc.

Source: *http://www.combimatrix.com/.*

✓ Related platform technology is commercially available.

Source: *http://www.azonano.com/details.asp?ArticleID=476.*

HORMONE THERAPY

Estrasorb

Novavax, Inc. (USA)

This is a topical lotion that contains estrogen "approved for the treatment of moderate to severe vasomotor symptoms (hot flashes) associated with menopause."

"The product utilizes Novavax's patented and proprietary micellar nanoparticle drug-delivery platform. This technology involves the application of a cosmetic-like moisturizing emulsion to deliver a therapeutic dose of 17 estradiol into the bloodstream when applied to the skin."

✓ Approved by the FDA in October 2003.

Source: *http://www.estrasorb.com/pdf/Esprit-ESTRASORB.pdf.*

IMMUNOSUPPRESSANT

Rapamune

Wyeth (USA)

An immunosuppressant "indicated for the prophylaxis of organ rejection in patients aged 13 years or older receiving renal transplants." (Data source: *http://www.wyeth.com.*)

The product uses Elan's nanocrystal technology to provide "more convenient administration and storage than Rapamune oral solution."

✓ Approved by the FDA in August 2000.

Source: *http://www.elan.com/edt/nanocrystal_technology/default.asp.*

OTHER AREAS

That's not all. Scientists are working now to create novel nanostructures that serve as new kinds of drugs for treating other afflictions, including Parkinson's and cardiovascular disease; to engineer nanomaterials for use as artificial tissues that would replace diseased kidneys and livers, and even repair nerve damage; and to integrate nanodevices with the nervous system to create implants that restore vision and hearing and build new prosthetic limbs.

1

Nanotechnology and Appetite Control

Anorexia nervosa is a common eating disorder, bemoaned on TV chat shows and fretted about in Hollywood gossip magazine articles about Angelina Jolie and other starlets. The disease is characterized by an inability, or sometimes a simple refusal, to maintain a healthy body weight, and a morbidly obsessive fear of gaining weight due to a false self-image.[1]

Anorexia may be maintained by various cognitive biases that alter how the affected individual evaluates and thinks about their body, food, and eating. It is a serious mental illness with a high incidence of comorbidity and also the highest mortality rate of any psychiatric disorder.[2]

Anorexia can affect men and women of all ages, races, socio-economics, and cultures.

The term *anorexia nervosa* was established in 1873 by Sir William Gull, one of Queen Victoria's personal physicians. The term is of Greek language origin: a (Α, prefix meaning in medicine, negation), n (ν, a link between two vowels) and orexis (ορεξισ, appetite), which basically means a lack of desire to eat.

A patient diagnosed with anorexia nervosa presents a number of symptoms. The severity of the symptoms vary in each patient's case and may present but not be readily apparent. Anorexia nervosa and the associated malnutrition that results from starvation can cause severe complications in every organ system in the body.

The major signs of anorexia include:

- Rapid, dramatic weight loss
- Scarring of the knuckles from placing fingers down the throat to induce vomiting
- Soft, fine hair growth on face and body
- Obsession with calories and fat content
- Morbid preoccupation with food, recipes, or cooking; may cook fancy dinners for others but not eat themselves
- Continuing to diet despite being dangerously underweight
- Obsessive fear of gaining weight or becoming overweight
- Demonstrates bizarre rituals: cuts food into tiny pieces; hides or discards food
- Purging using laxatives, diet pills, ipecac syrup, or water pills
- Induces vomiting; may scurry to the bathroom after eating to vomit and purge themselves of the calories
- Engages in frequent, strenuous exercise
- Perceives self to be overweight despite being told by others they are too thin

- Becomes intolerant of cold weather and constantly complains of feeling cold due to loss of insulating body fat; body temperature lowers (hypothermia) in effort to conserve calories
- Depression, and may frequently be in a sad, lethargic state
- Solitude: may avoid friends and family; becomes withdrawn and secretive
- Dons baggy, loose-fitting clothes to cover weight loss
- Cheeks may become swollen due to enlargement of the salivary glands caused by excessive vomiting

Cultural studies have highlighted the role of social factors, such as the promotion of thinness as the ideal female form in Western industrialized nations, particularly through the media.[3]

A recent epidemiological research project involving a total of 989,871 Swedish men and women indicated that gender, ethnicity, and socioeconomic status were huge influences on the chance of developing anorexia. Those with non-European parents were among the least likely to be diagnosed with the condition, and those in wealthy, white families were at the greatest risk.

Men and women in professions where there is a social pressure to be thin—including actors, models, and dancers—were more likely to develop anorexia during the course of their lives, and further studies have suggested that those suffering from anorexia may have higher contact with cultural influences that promote weight loss.

Right now, the primary forms of medical therapy for this disease are talk therapies, including:

- Acceptance and commitment therapy: A type of therapy that has shown promise in the treatment of anorexia in which participants show clinically significant improvement on at least some measures; no participants worsened or lost weight even at one-year follow-up point.
- Cognitive remediation therapy (CRT): This is a cognitive rehabilitation therapy, developed at King's College in London, designed to improve such neurocognitive abilities as attention, working memory, cognitive flexibility and planning, and executive functioning which leads to improved social functioning. Neuropsychological studies have shown that patients with anorexia have difficulties in cognitive flexibility. In studies conducted at King's College and in Poland with adolescents, CRT was proven to be beneficial in treating anorexia nervosa; in the United States, clinical trials are still being conducted by the National Institute of Mental Health on adolescents age 10 to 17, and at Stanford University with subjects over 16, as a conjunctive therapy with cognitive behavioral therapy, according to the researchers involved.
- Family therapy: Various forms of family therapy have been proven to work in the treatment of adolescent anorexia nervosa including *conjoint family therapy (CFT),* in which the parents and child are seen together by the same therapist, and *separated family therapy (SFT),* in which parents and child attend therapy separately with different therapists.

Some doctors have tried antidepressants, but these drugs treat only a symptom, not the cause, of the disease.

Anorexia may evolve into to a more severe condition, which is called *cachexia*. Cachexia is a physical wasting with loss of weight and muscle mass in someone not actively trying to lose weight. In cachectic patients, the metabolic, or energy, demands of the body are often increased.

When the human body does not obtain sufficient calories and nutrients to meet its energy requirements, it uses fat and muscle as a source of energy. This results in weight loss and severe wasting. That's where medical nanotechnology innovation comes in to play.

A firm called Par Pharmaceutical Companies, Inc. (USA), through one of its operating units, Strativa Pharmaceuticals, has developed a medical nanotechnology called Megace ES. This nanotechnology-based drug is designed to stimulate the appetite for "treatment of anorexia, cachexia, or an unexplained, significant weight loss in patients." The drug is primarily for those with a diagnosis of acquired immunodeficiency syndrome (AIDS), but newer research is pointing to the possible efficacy of the drug for others who suffer from anorexia and related disorders.

The technology utilizes a nanocrystal technology delivery system to improve the rate of dissolution and bioavailability of the original megestrol acetate oral suspension.

Nanocrystal particles are small particles of drug substance, typically less than 1000 nm in diameter, which are produced by milling the drug substance using a proprietary, wet-milling technique. The nanocrystal particles of the drug are stabilized against agglomeration by surface adsorption of selected GRAS stabilizers. The result is an aqueous dispersion of the drug substance that behaves like a solution—a nanocrystal colloidal dispersion, which can be processed into finished dosage forms for all routes of administration. The drug was approved by the FDA in July 2004.

Taste, Technology

"What does it taste like," patients commonly ask? Megace ES is presented as a milky-white, lemon-lime flavored, concentrated oral suspension.

This nanotechnology-based drug is basically a reformulation of megestrol acetate oral suspension. Research has shown that Megace ES has demonstrated improved bioavailability for subjects in the unfed state versus original formula Megace®.

For AIDS patients who have no interest in eating or are seeing extreme weight loss, Megace ES can help improve appetite and gain weight, which can lead to an enhanced sense of well-being, researchers report.

Nanotechnology-based Megace ES is innovative and is decidedly different from original formula Megace, as it is:

- More easily absorbed.
- Has improved bioavailability in the unfed condition versus original formula Megace and equivalent bioavailability in the fed condition.
- Achieves the fed condition in subjects eating a high-calorie (800 to 1000 calories) and a high-fat meal.

- Has 75 percent less volume and 94 percent reduced viscosity, as compared with Megace 800 mg/20 mL. This is vital for patients who are battling with themselves over if, and when, to eat. This also makes the drug easier to swallow for those who have difficulty with that process, as a comorbidity.
- Has less viscosity than commonly used over-the-counter (OTC) products such as Mylanta.

The patent protection obtained by developer Par Pharmaceuticals ensures that, for the coming decade, there will be no generic substitute for Megace ES. The nanocrystal technology is the unique differentiator that helped the company obtain the patent for the drug. The nanocrystal particles used in the drug are up to 50 times smaller than the micronized particles in the original formula of Megace. These tiny particles are stabilized and enable an aqueous dispersion of the drug. The nutritional value of the standard dose has been compared by researchers to one of the Food and Drug Administration's (FDA's) "example meals" for breakfast (Fig. 1.1).

FIGURE 1.1 FDA model breakfast, courtesy of Par Pharmaceuticals.

This "example meal" consists of two eggs fried in butter, two strips of bacon, two slices of toast with butter, four ounces of hash brown potatoes, and four ounces of whole milk.

As averred to above, the weight gain associated with Megace ES and was observed in patients with HIV-associated unintended weight loss during a pilot study. See Fig. 1.2.

The trial referred to above was conducted as a pilot study using a concentrated suspension of 575 mg/5 mL megestrol acetate. Megace ES is offered as megestrol acetate 125 mg/mL oral suspension. The recommended dose of Megace ES is 625 mg/5 mL. The randomized, open-label, multicenter pilot study enrolled 63 AIDS patients, and these patients received either Megace ES (575 mg/5 mL) or Megace (800 mg/20 mL) once every morning for 12 weeks.

The results were rather impressive. Those patients who received megestrol acetate oral suspension 800 mg/20 mL

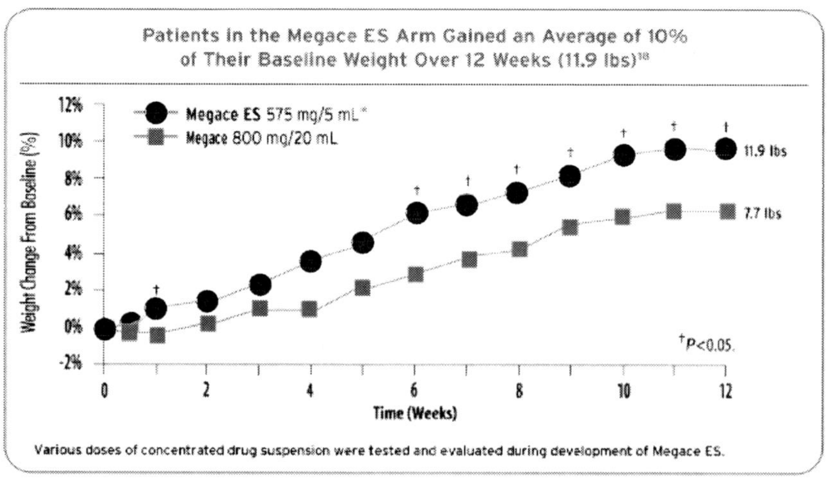

FIGURE 1.2 Chart courtesy of Par Pharmaceuticals.

gained an average of 6 percent of their baseline weight over 12 weeks, or 7.7 lb, according to researchers.

Remember: unintended weight loss can be a common disorder in patients with HIV/AIDS. For these patients, maintaining respectable caloric intake is extremely difficult.

Unintended weight loss can greatly impact AIDS patients in the following ways:

- Loss of physical functions
- Increased mortality risk
- Significant risk of hospitalization
- Depression, related mood disorders, and withdrawal from daily activities

To be sure, unexpected weight loss is associated with morbidity and a reduced quality of life.

Even a minimal weight loss of 5 percent is a predictor of opportunistic infections in the AIDS patient and can lead to anorexia or cachexia. The safety results were similar for each branch of the trial, according to researchers.

Figure 1.3 provides further details on how patients increased their caloric intake, in addition to gaining weight, and demonstrating an increased appetite.

Most compellingly for this nanotechnology-based drug, 89 percent of patients showed an improvement in appetitive versus just 50 percent of patients on the placebo. Patients also demonstrated an "increased sense of well-being" while taking the drug during the trial, according to the researchers. This included improved, positive views of their own health, lessened concern

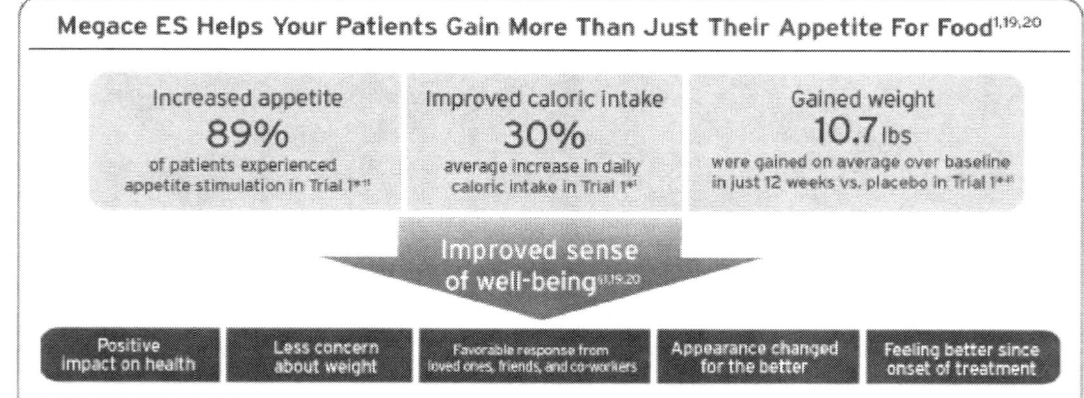

FIGURE 1.3 Chart courtesy of Par Pharmaceuticals.

about their weight, an improved response from family members and other loved ones, and even improved appearance.

A second trial demonstrated strong results for this nanotechnology innovation as well. The second trial included 65 patients, and there was a mean decrease of 1.7 lb in lean body mass during the trial period and a mean increase in weight of 5.7 lb in body mass from the baseline. Those patients on the placebo during the second trial demonstrated much less weight gain—only 1.5 lb in mass from the baseline.

Side Effects, Safety Concerns

No drug is without side effects, including this nanotechnology drug for appetite control. According to the developers, the most prevalent adverse events linked to Megace ES 625 mg/5 mL and megestrol acetate suspension 800 mg/20 mL include impotence, flatulence, rash, hypertension, fever, reduced libido, insomnia, dyspepsia, and hyperglycemia. Females who participated in the clinical trial reported breakthrough bleeding.

Megace ES and megestrol acetate oral suspension are contraindicated in patients with a history of hypersensitivity to megestrol acetate or any component of the formulation, or in females known to be pregnant. Women who are parents of newborns should discontinue nursing while on the drug, the researchers indicate.

There are other side effects, too—new-onset diabetes mellitus, aggravation of pre-existing diabetes mellitus, Cushing's syndrome, and insufficiency of adrenaline.

Researchers also indicate that doctors should be cautious in using the nanotechnology drug in treating patients with thrombo-embolic disease. Physicians may need to modify the dosing for these patients as well as for the elderly and those with chronic renal disease.

Adverse events for the drug are greater than the placebo, i.e., ≥5 percent, researchers said.

On a related note, with a daily dosing of just 1 tsp once daily, Megace ES is seen to be more convenient for patients compared with original formula Megace, which has a dose of 4 tsp once daily.

As suggested above, original formula Megace has a consistency similar to Mylanta, while Megace ES has a consistency similar to whole milk, which may make Megace ES easier to swallow for some (see Fig. 1.4).

There is much general concern in medicine about the toxicology associated with drugs based on nanotechnologies. One recent report[4] was published by authors from the Education and Information Division, National Institute for Occupational Safety and Health, Cincinnati, OH. According to these researchers, nanotechnology works because of the simple fact that nanoparticles demonstrate exclusive physicochemical properties, which are quite distinct from fine-sized particles of a similar composition. The researchers said that, accordingly, nanoparticles may also demonstrate distinct bioactivity and innovative interactions with biological systems.

Thus, it is vital to "assess the potential health risks of exposure to nanoparticles to allow development and implementation

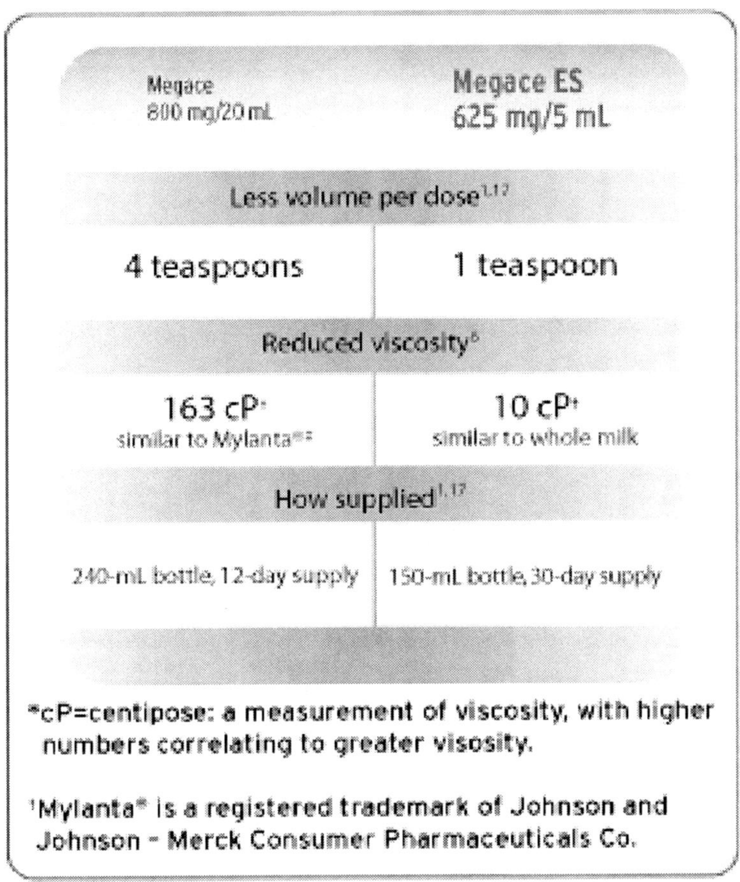

FIGURE 1.4 Chart courtesy of Par Pharmaceuticals.

of prevention measures," the authors wrote. "Risk assessment requires data concerning hazard and exposure."

However, several challenges are quite daunting for the field of nanotoxicology, especially obtaining the required facts for assessment of the bioactivity of nanoparticles. These include, the authors note, (a) the large number of nanoparticle types to be evaluated, (b) the desire to use nanoparticle doses and structure sizes in cellular and animal test systems, and (c) *in vitro* results that show the actual absorption of nutrients or assay indicator compounds from the culture media. This opinion piece, to be sure, challenges some of the views of the progress made in the field of nanotoxicology in recent years to overcome these challenges.

Other researchers have made similar points recently. An article called "Nanotechnology, Risk, and Oversight: Learning Lessons from Related Emerging Technologies"[5] was published by researchers at the Hubert H. Humphrey Institute of Public Affairs, University of Minnesota, and the School of Mass Communications, University of Nevada–Las Vegas.

According to these researchers, in general, emerging technologies such as nanotechnology in medicine are utilized primarily for their novelty and therefore are marked by "significant uncertainty" in discerning the correct way to manage associated risks.

"There is a body of prior knowledge about risk management and oversight policy for other technologies that have already permeated society," the authors note. "Here, we describe two ways in which prospective oversight policy analysis for emerging technologies can draw upon these past experiences."

One such method involves comparing nanotech products that have already been marketed to similar products of the emerging technology—the so-called *cognate-product approach*. The other method examines an emerging technology as a genus of products and then compares it to another technological field that has already emerged and penetrated markets—the so-called *whole-technology approach*.

There are "parallels between biotechnology and nanotechnology as whole fields of development and also between genetically engineered organisms in the food supply and agricultural products of nanotechnology," the authors note. "We find that both approaches to historical learning have value and present lessons that could be applied to nanotechnology."

To address these growing concerns, researchers are developing computational strategies to predict the risks associated with nanotechnology drugs.

One recent article on this trend appeared in *Nanoscale*, called "Computational strategies for predicting the potential risks associated with nanotechnology."[6] The author is with Australia's Commonwealth Scientific and Industrial Research Organization (CSIRO) Materials Science and Engineering and Future Manufacturing Flagship. The researcher notes that for the "move from nanoscience to nanotechnology to be sustainable," it is vital that the issues surrounding possible nano-hazards be handled before commercialization.

The global push for more environmentally friendly, biodegradable products means that the introduction of the nanoparticles contained within these products into the ecosystem is an inevitability. When this happens, it is desirable to know how the hazardous properties will affect us and what potential hazards exist.

There are worries that the regulations for nanotech are insufficient. One recent article on this topic was titled, "A global view of regulations affecting nanomaterials."[7] The author was S. F. Hansen, from the Department of Environmental Engineering, Technical University of Denmark (Kongens Lyngby 2800, Denmark). According to the author, the decade of the 2000s was characterized "by an unprecedented exploration into research and development" of nanotechnology and nanomaterials.

"Despite a slow start, new regulatory initiatives are popping up like mushrooms internationally. Many of these initiatives have yet to materialize or are soft law* initiatives, and their impact on the development of more authoritative and prescriptive regulatory measures is most likely to be limited," the author noted. "This is due to a number of transnational regulatory challenges that include (a) whether to adapt existing legislation or develop a new regulatory framework, (b) whether nanomaterials should be considered as different from their bulk counterparts, (c) how to define nanotechnology and nanomaterials, and (d) how to deal with the profound limitations of risk assessment when it comes to nanomaterials. In this opinion, I discuss these and related issues and conclude that the development of a new authoritative and prescriptive regulatory framework might be the only way to effectively address these challenges while ensuring a transparent and informed decision-making process."

But, contrary research is available as well.

A piece entitled, "Safety and Efficacy of Two Preparations of Megestrol Acetate in HIV-Infected Individuals with Weight Loss in Africa, India, and the United States," appeared in the *Journal of Applied Research* in 2007.[8] The authors were from the Department of Medicine and Public Health at Tufts University and at Par Pharmaceuticals.

---

* i.e., quasi-legal instruments that have little or no legally binding force.

The authors argue that nanotechnology drugs for appetite control, such as Megace ES, were safe and effective, and demonstrated no known toxicities for patients. "Megestrol acetate improved body mass index (BMI); weight gain was significantly more rapid and substantial with the concentrated suspension," the authors wrote.

The unknown toxicity risks, at this point, outweigh the benefits for patients suffering from appetite control issues. "Weight loss remains a frequent and troubling complication for patients," the authors wrote. "Weight loss remains a predictor of mortality."

They also noted—as averred to above—that weight loss also contributes significantly to other comorbidities, like reduced quality of life.

"The use of an appetite stimulant such as megestrol acetate has been shown to dramatically increase caloric intake in patients," the authors note.

Developers of the drug took a number of precautions to ensure safety and efficacy, per FDA regulations.

"Safety monitoring included measures of serum cortisol, and adrenocorticotropic hormone (ACTH) stimulation testing," the authors wrote. "There was also testing of liver function, lipid panel, fasting glucose, and hemoglobin A1c (HbA1c) prior to and after the trial." The trial of the drug, as averred to above, was 12 weeks. "Assessment of efficacy included monitoring of patients' dietary intake and body weight and composition during the trial."

Weight of patients was measured when they were dressed in street clothes, without shoes, and anthropometry, including mid-arm, waist, and hip circumference as well as triceps skin fold measurement, the researchers reported. "Body composition was assessed by bioelectrical impedance analysis (BIA) performed on the Quantum II analyzer (RJL Systems, Clinton, Mich.), with included software (Cyprus 2.7; RJL Systems)."

Lab samples that were obtained were sent to a central laboratory for processing of CBC, WBC, platelets, HbA1c, sodium, potassium, chloride, bicarbonate, BUN, albumin, glucose, creatinine, alkaline phosphatease, total bilirubin, AST/ALT, LDH, lipid panel, and urinalysis.

"Subjects were stratified by country and randomized from a central site," the authors noted. "Study visits were conducted weekly to monitor safety."

Assessments of weight, anthropometry, and BACRI were reported weekly, the researchers noted. "Thirty days after the completion of the study drug, subjects were seen for assessments of safety," wrote the authors. "Subjects returned for additional laboratory tests, as needed."

What is more, efficacy analyses were performed on the intention-to-treat basis of all randomized subjected who received medicine. "Safety summaries were performed on all subjects who received at least one dose of medication," the researchers wrote. "All analyses were performed using SAS statistical software, version 8.2."

For the study, a total of 121 patients were screened. "We have been able to demonstrate that use of an appetite stimulant could safely encourage dietary intake and result in gain of both lean and fat weight, suggesting that interventions such as this will be useful," wrote the authors. "Formulations of megestrol acetate were demonstrated to be safe and successful."

Availability

Megace ES is now Tier 2 (preferred brand-name drug) on Silver Script/Rx America Part D, as of the spring of 2010—i.e., Medicare's Part D plans.

The developer, Par Pharmaceuticals, through its operating unit, Strativa Pharmaceuticals, is telling health care practitioners that if their patients don't have prescription coverage through their health insurance carrier, and, as a consequence, and cannot afford Megace ES, the company is offering RxHope, a patient assistance program.

Patients, physicians, and patient advocates can apply for eligible, needy patients. Patients are eligible if they do not have any prescription coverage for Megace ES through any private or government funded prescription programs, including Medicare, Medicaid, and Medicare Part D. Go to www.megacees.com/ for more information.

References

[1] Rosen, J. C., J. Reiter, and P. Orosan (Jan. 1995). Assessment of body image in eating disorders with the body dysmorphic disorder examination. *Behav Res Ther* 33(1): 77–84. doi: 10.1016/0005–7967(94)E0030-M. PMID 7872941.

[2] Attia, E. (Feb. 2010). Anorexia Nervosa: Current Status and Future Directions, *Annu Rev Med* 61: 425–35. doi: 10.1146/annurev.med.050208.200745. PMID 19719398.

[3] Media Harming People's Body Image. nursing-resource.com (accessed Sept. 1, 2010).

[4] Geraci, C. L., and V. Castranova. Challenges in assessing nanomaterial toxicology; a personal perspective. *Wiley Interdiscip Rev Nanomed Nanobiotechnol* 2010 Nov.–Dec. 2(6): 569–77. PMID: 20799267 [PubMed—in process].

[5] Kuzma, J., and S. Priest. Nanotechnology, Risk, and Oversight: Learning Lessons from Related Emerging Technologies. *Risk Anal* Aug. 17, 2010. PMID: 20723152.

[6] Barnard, A. S. Computational Strategies for Predicting the Potential Risks Associated with Nanotechnology. *Nanoscale* Oct. 21, 2009, 1(1):89–95. Epub Aug. 13, 2009.

[7] Hansen, S. F. A Global View of Regulations Affecting Nanomaterials. *Wiley Interdiscip Rev Nanomed Nanobiotechnol* Sep.–Oct. 2010, 2(5):441–9. PMID: 20533518.

[8] Wanke, C., MD; J. Gutierrez, BA; A. Kristensen, MS; and L. MacEarchen, PHD, MPH. Safety and Efficacy of Two Preparations of Megestrol Acetate in HIV-Infected Individuals with Weight Loss in Africa, India, and the United States. *J Appl Res* Vol. 7, No. 3, 2007.

Suggested Reading

Geraci, C. L., and V. Castranova. Challenges in assessing nanomaterial toxicology; a personal perspective. *Wiley Interdis-*

cip Rev Nanomed Nanobiotechnol 2010 Nov.–Dec. 2(6): 569–77.

Wanke, Christine, MD; Jodi Gutierrez, BA; Allan Kristensen, MS; and Laureen MacEarchern, PHD, MPH (Dept. of Medicine and Public Health at Tufts University and at Par Pharmaceuticals.) Safety and Efficacy of Two Preparations of Megestrol Acetate in HIV-Infected Individuals with Weight Loss in Africa, India, and the United States. *J Appl Res* Vol. 7, No. 3, 2007.

Hansen, S. F. (Department of Environmental Engineering, Technical University of Denmark, Kgs. Lyngby 2800, Denmark). A global view of regulations affecting nanomaterials. *Wiley Interdiscip Rev Nanomed Nanobiotechnol* 2010 Sep.–Oct. 2(5):441–9.

Barnard, A. S. (Commonwealth Scientific and Industrial Research Organization Materials Science and Engineering & Future Manufacturing Flagship Clayton, Victoria, Australia). Computational strategies for predicting the potential risks associated with nanotechnology. *Nanoscale* 2009 Oct. 21 1(1):89–95. Epub 2009 Aug. 13.

2

Nanotechnology and Cancer Therapy

Tracking, Targeting Tumors

Lonnie Moulder shouldered his way to the front of the room, cleared his throat, and gave his best pitch to savvy investors at the New York Palace Hotel in New York City. The date—May 13, 2009. The time—exactly 2:40 p.m. Moulder, the president and CEO of Abraxis BioScience, told a compelling sales story about the innovative use of nanotechnology in cancer therapy, at the Merrill Lynch 2009 Health Care Conference. Abraxis was trading on the NASDAQ Global Market under the symbol ABII at the time. Interest in the sales presentation was intense—and was even broadcast over a webcast to investment bankers in Hong Kong, London, and Tokyo, and to doctors at hospitals all over the U.S.A. and Europe.

The investors—many of whom had been following the emerging firm for a few years already—heard news about the latest developments at Abraxis BioScience, developer of what it called the "next generation" of therapeutics for cancer patients. The firm's portfolio of products included the first patented, protein-bound, nanoparticle chemotherapeutic compound, called Abraxane. Moulder briefed the investors about the company's proprietary tumor-targeting technology known as the "nab® platform."

As a result of all the hype, carefully cultivated by Moulder, Celgene acquired Abraxis for $2.9 billion in cash and stock. The clinical success of the world's first nanotechnology cancer treatment—the focus of Moulder's talk that day in 2009—had led to stock market success, too. The transaction gave Celgene instant commercial operations in the solid tumor market. The pharma company reckons that it could add $1 billion to its annual sales by 2015. Abraxane brought in about $315 million in sales in 2009, 88 percent of Abraxis' revenue.

The Food and Drug Administration (FDA) first approved Abraxane in 2005, for treatment of metastatic breast cancer. Moulder told the investors what they wanted to know about the company's growth prospects—the therapeutic had now been approved by regulators in 40 countries. The company, he pledged to investors and physicians, was expanding its platform through a clinical program and had many new nanotechnology cancer therapies "in its deep product pipeline."

The drug Moulder discussed was an albumin-bound form of paclitaxel with a mean particle size of approximately 130 nm.

As Moulder mentioned to investors on that spring day in 2009, Abraxis BioScience was moving forward with other new applications for the nanotechnology drug. The new apps were the key to future growth. Although originally approved for treatment of breast cancer, the company secured orphan drug status for a version of the drug for treatment of pancreatic cancer.

The latest data from the company demonstrated that Abraxane has "potential" for treatment of patients with stage III–IV melanoma.

This kind of cancer is known to medical science as being one of the toughest cancers to treat. But new data, released in the summer of 2010, around the time Celgene acquired Abraxis, showed that many new studies evaluated the drug Abraxane for advanced melanoma therapy and other cancers, too. The scientific evidence was overwhelming—Abraxis BioScience published an amazing 41 scientific abstracts at the 46th Annual Meeting of the American Society of Clinical Oncology laying out the scientific story.

"The outlook for Abraxis is immensely encouraging," said founder and Chairman Dr. Patrick Soon-Shiong, who is still with the firm, now a division of Celgene.

Independent medical literature confirmed this hypothesis, according to a report in *the International Journal of Nanomedicine.*[1] Researchers noted that breast cancer is the most prevalent type of cancer diagnosed in women. During the metastatic phase, this disease is incurable, according to leading oncologists.

Taxanes, developed in recent years, are an important class of antitumor agents which are fundamental in the treatment of advanced and early-stage breast cancer. However, clinical advances in the development of taxanes have been restricted by their highly "hydrophobic molecular status," the researchers noted. To resolve this poor water solubility, lipid-based solvents have been employed by scientists as a vehicle, and a number of systemic formulations have been created, primarily for paclitaxel, which are Cremophor-free and thus foster the circulation of the drug.

But nanotechnology-based ABI-007 is a novel, albumin-bound, 130-nm particle formulation of paclitaxel, which has no solvents. Research conducted by the Italians showed this nanotechnology formulation to be superior to "an equitoxic dose of standard paclitaxel" with a much smaller incidence of toxicities. This was found during a large, international, randomized Phase III trial. The researchers indicate that the availability of this new drug, used in association with other traditional and nontraditional drugs (like the brand new antineoplastic agents and so-called targeted molecules) give the oncologist an array of effective treatment options for patients.

Breast Cancer Backgrounder

More than 180,000 estimated new cases of metastatic breast cancer were diagnosed in the U.S.A. in 2008, making it one of the most common forms of cancer here. Nearly one third—or 32 percent—of all cancers diagnosed in women are breast cancer related.

"In the metastatic setting, this disease is as yet incurable," the Italian researchers report. "And the main objectives are the palliative prolongation of survival and improvement of quality of life."

Other therapeutic end points are (a) response rate, (b) time to progression, (c) time to treatment failure, and (d) others.

"All of these are surrogate end points without any real advantage for the patients suffering from a metastatic and progressive disease," the Italian researchers note.

Thus, breast cancer persists as a significant health problem, although the mortality rate has decreased due to the new treatments of new drugs that have shown demonstrable advantages for women, including with advanced disease.

According to the Italian researchers, taxanes, like paclitaxel (Taxol®; Bristol-Myers Squibb Co., Princeton, N.J.) and docetaxel (Taxotere®; Aventis Pharmaceuticals Inc., Bridgewater, N.J.) are a new class of antitumor agents that are assisting in the treatment of advanced and even early-stage breast cancer.

These two drugs are included in the now-common treatment regimens for adjuvant chemotherapy and are the preferred agents for recurrent and metastatic breast cancer by the *National Comprehensive Cancer Network (NCCN) Clinical Practice Guidelines.*

These taxanes are cell cycle-specific targeting agents that bind with microtubules, thus stabilizing tubulin polymerization and tamping down spindle microtubule dynamics.

The drugs effectively inhibit mitosis, motility, and intracellular transport within cancerous cells. That leads to apoptotic cell death. What is more, these drugs have demonstrated antineoplastic activity against a wide variety of malignancies including also non-small-cell lung cancer and even cancer of the ovaries.

Interestingly, paclitaxel is derived from a naturally occurring complex product that is extracted from the bark of the western yew tree, or *Taxus brevifolia.* Docetaxel was originally isolated in a precursor form from the needles of the European yew tree.

Although these clinical advances are impressive, there are limits to the effectiveness. The chemical formulation of taxanes is characterized by highly hydrophobic molecules.

"To overcome this poor water solubility, lipid-based solvents are used as a vehicle," the Italian researchers noted.

Researchers say that the solubility of paclitaxel is generally bolstered with a mix of 50:50 Cremophor EL®—i.e., CrEL, a non-ionic surfactant polyoxyethylated castor oil (BASF, Florham Park, NJ) and ethanol. Docetaxel is created with polysorbate 80—Tween® 80—and ethanol diluents like Taxotere.

To administer the drugs, both drugs are further diluted from 5- to 20-fold with conventional saline or 5 percent dextrose solutions before IV infusion.

"These solvent-based formulations, however, can be associated with serious and dose-limiting toxicities," the Italian authors noted.

According to the Italian researchers, polyoxyethylated castor oil is both biologically and pharmacologically active. This combination leaches plasticizers from standard intravenous (IV) tubing releasing di-, or 2-ethylhexyl-phthalate (DEHP).

"Its infusion produces histamine release with consequent well described hypersensitivity reactions, including anaphylaxis," the Italian researchers noted.

According to the doctors who wrote the above-cited report, in early Phase I trials, 20 percent to 40 percent of non-premedicated patients were affected by these adverse reactions. What is more, it has been also associated with hyperlipidemia, abnormal lipoprotein patterns, aggregation of erythrocytes, and prolonged, sometimes irreversible sensory neuropathy, which may be associated with demyelination and axonal degeneration. "CrEL can also cause neutropenia," the authors noted.

Physicians noted that the CrEL–paclitaxel formulation requires special infusion sets—i.e., tubing and in-line filters—to minimize exposure to DEHP.

On the other hand, longer infusion times—1 to 24 hr, median time 3 hr—with a large volume of IV fluid and premedication, including dexamethasone, diphenhydramine, and cimetidin, help to reduce the risk of hypersensitivity reactions. Despite these standard precautions, hypersensitivity can, however, occur and rarely be fatal.

"Infusion schedules of paclitaxel seem to influence its clinical effectiveness, too. In fact, longer infusions of the drug produce greater clinical efficacy than more rapid injections," the researchers wrote.

What is more, hypersensitivity reactions also occur with polysorbate 80, but to a lesser extent than with CrEL. Additionally, Polysorbate 80 has been linked with severe and irreversible sensory and motor neuropathies.

Moreover, polysorbate 80 may alter membrane fluidity, causing cumulative fluid retention. This unique docetaxel toxicity may be reduced by prophylactic corticosteroids.

Researchers also note the most damaging side effect: CrEL and polysorbate 80 may limit tumor penetration with a negative impact on efficacy. The formation of large polar micelles of CrEL–paclitaxel in the plasma compartment traps the drug and can lead to nonlinear pharmacokinetics and decrease drug clearance and decrease volume of distribution.

"This contributes to a lack of dose-dependent antitumor activity," wrote the authors. "Also coadministered drugs, such as anthracycline compounds, could be affected by this phenomenon."

Lastly, it has been recently shown that *in vitro* polyoxyethylated castor oil truly inhibits endothelial transcytosis of paclitaxel that is mediated by an albumin receptor.

"During recent years, a special effort has been made to avoid these problems," the Italian researchers indicated. "New systemic formulations are being developed, mostly for paclitaxel, which are highly soluble, Cremophor-free, and increase the circulation time of the drug."

Impact of Nanotechnology Developments on the Perceptions of Oncologists

According to the researchers, cancer doctors looked to nanotechnology for help in solving these seemingly insoluble problems. Nanotechnology is an interdisciplinary field, just like oncology, that over the last few years has helped medicine overcome many therapeutic problems. Remember that nanotechnology refers to small-scale, applicable materials, which are 1 to

100 nm in size. Nanotechnologies are characterized by (a) the size of the device, (b) its man-made characteristics, (c) other properties linked to its nanoscopic dimensions, and (d) the ability of ad hoc mathematical models to predict specific behaviors of nanoscale materials.

Cancer researchers first looked to nanotechnologies as a solution after seeing what was done with the technologies with liposomes, dendrimers, super paramagnetic nanoparticulates, polymer-based platforms, gold nanoshells, silicon- and silica-based nanoparticles, carbon-60 fullerenes, and nanocrystals.

"They can be divided into three generations of compound, according to whether or not they were developed to target a specific target which is expressed on the tumor cells or the endothelium," the Italian authors noted.

"Among the first generation's vectors—not specifically targeted—liposomal drug delivery is certainly the most successfully used in the clinic, as demonstrated by liposomal doxorubicin for breast, ovarian, and Kaposi's sarcoma," the authors wrote.

That fact caught the eye of oncology researchers. They figured that albumin had a number of characteristics that made it a likely effective drug delivery vehicle for oncology, as it is a natural carrier of endogenous, hydrophobic molecules such as vitamins, hormones, and other water-insoluble biologics, like plasma. All of those molecules, moreover, are bound in a reversible, noncovalent manner.

What is more, albumin helps endothelial transcytosis of protein-bound and unbound plasma constituents primarily through binding to a cell surface; 60-kDa glycoprotein receptor, albondin, gp60 binds to caveolin-1, an intracellular protein with subsequent formation of transcytotic vesicles, caveolae, according to the Italian researchers.

Additionally, osteonectin, a secreted protein acid rich in cysteine, has been shown to bind albumin because of a sequence homology with gp60. SPARC (for *secreted protein acidic and rich in cysteine*), also known as caveolin-1, is often present in lung, breast, and prostate neoplasms, which may be the explanation for why albumin is known to accumulate in some tumors and thus facilitates intratumor accumulation of albumin-bound drugs.

That's why albumin-bound nab-paclitaxel ABI-007 was developed and was embraced enthusiastically by physicians in oncology. Abraxane® is, of course, made by Abraxis BioScience, but AstraZeneca has some intellectual property rights over the drug too.

"This nanovector application for breast cancer represents one of the strategies adopted to overcome the solvent-related problems of paclitaxel, and it has been recently approved by the U.S. Food and Drug Administration for pretreated metastatic breast cancer patients," the researchers wrote.

Researchers note that Abraxane, ABI-007, is a novel, albumin-bound, 130-nm particle formulation of paclitaxel, free from any solvent. The drug is "used as a colloidal suspension derived from the lyophilized formulation of paclitaxel and human serum albumin diluted in saline solution, 0.9 percent NaCl," the Italian researchers reported.

Human serum albumin helps stabilize the drug particle at an average size of 130 nm, which dramatically reduces the risk of capillary obstruction and does not necessitate any particular infusion systems or steroid/antihistamine premedication before the infusion, the researchers report.

According to a search of the medical literature, preclinical studies in athymic mice with human breast cancer showed that ABI-007 has a “higher penetration” into tumor cells with an increased anti-tumor activity, compared with an equal dose of standard paclitaxel.

One Phase I clinical study by Ibrahim, conducted with 19 patients who had solid tumors and breast cancer, demonstrated a maximum tolerated dose of ABI-007 about 70 percent higher than that of CrEL paclitaxel formulation, i.e., 300 mg/m^2 for a regimen of three times a week.

To be sure, there were still dose-limiting toxicities. Side effects included sensory neuropathy, stomatitis, and ocular toxicity, i.e., superficial keratopathy and blurred vision at a dose of 375 mg/m^2.

But, importantly, not one patient experienced a hypersensitivity reaction. The drug ABI-007 was administered intravenously with no premedication, in shorter infusion periods, i.e., 30 min versus 3 hr for polyoxyethylated castor oil-based paclitaxel, and with a standard infusion device. What is more, pharmacokinetic parameters showed a decidedly linear trend.

Then, another study, a Phase II trial, demonstrated that ABI-007 has “important antitumor activity” in patients with metastatic breast cancer, and that drew even more attention to the nanotechnology drug.

The overall response rate, at a dose of 300 mg/m^2 every 3 weeks, was 48 percent for all patients and 64 percent for patients in “first-line” therapy, the Italian researchers reported.

“Time to tumor progression was 26.6 weeks for all patients and 48.1 weeks for patients with confirmed tumor responses; median overall survival was 63.6 weeks. No severe ocular events were noted, and other common taxanes-associated toxicities were less frequent and less severe, e.g., myelosuppression, peripheral neuropathy, nausea, vomiting, fatigue, arthralgia, myalgia, alopecia,” the Italian researchers noted.

Lastly, in a large international randomized Phase III study, equitoxic doses of ABI-007, 260 mg/m^2, and polyoxyethylated castor oil-based paclitaxel, 175 mg/m^2, were compared in 454 patients with metastatic breast cancer.

“ABI-007 was superior to standard paclitaxel for both overall response rate—33 percent versus 19 percentage, respectively; $p = 0.001$, and time to tumor progression, $p = 0.006$, in all subgroups of patients, but mostly for those receiving the drug as first-line therapy, 42 percent versus 27 percent, respectively; $p = 0.029$, the Italian researchers found.

“Also in this trial, the incidence of toxicities was significantly lower in the ABI-007 group than the polyoxyethylated castor oil-based paclitaxel group; in particular, grade 4 neutropenia was lower, 10 percent versus 21 percent, respectively; $p = 0.001$, despite the approximately 50 percent higher dose,” the researchers noted.

Grade three sensory neuropathy was more frequent in the ABI-007 group, 10 percent versus 2 percent, respectively; p = 0.001. "But it was easily managed," the researchers said.

This is when the medical professional really began to notice the new nanotech drug, which increased the increased antitumor activity of ABI-007 via higher intratumor paclitaxel concentrations—as reported in preclinical studies—and a higher administered dose.

"Neymann et al. demonstrated also that weekly dosing of ABI-007 is safe and produces minimum toxic adverse effect with objective antitumor responses in patients previously exposed to paclitaxel," the researchers reported. They noted that further laboratory and clinical research is needed to advance the therapy of patients with breast cancer so as to improve the therapeutic index and the safety of this class of agents.

"In this context, the knowledge and investigation of new taxanes constitutes today one of the most exciting strategies for improving the clinical control of breast cancer in association with other cytotoxic agents not usually used in this disease, i.e., cisplatin, carboplatin, and irinotecan, and molecularly targeted agents, such as inhibitors of epidermal growth factor receptor, angiogenesis," the Italian researchers wrote.

"The formulation of Abraxane®, paclitaxel is delivered in a suspension of albumin particles, showing significant advantages to paclitaxel and its generic equivalents, in which polyethoxylated castor oil, Cremophor EL, is used as the solvent.

"These effects were achieved without increased toxicity to normal tissues, though the drug dose was 1.5 times higher than the maximum tolerated dose of traditional paclitaxel," the Italian researchers reported. What is more, the researchers showed, Abraxane has demonstrated a strong antitumor activity when associated with radiotherapy in a supra-additive manner.

According to all of the data cited above, there is powerful evidence that the association of Abraxane® with radiotherapy would further "improve the clinical results" of taxane-based chemoradiotherapy.

"In our opinion, in the near future this new taxane should be tested in large randomized clinical chemoradiotherapy trials," they concluded. "It is evident that treatment for breast cancer in future should be tailored to each patient, trying to select treatment strategies for cancer individually based on tumor-expressing factors and/or genomic and proteomic analysis. On the other hand, treatment strategies should be based not only on prognostic and predictive factors and prior adjuvant chemotherapy, but also on safety profile, impact on quality of life, and patient preference. We believe that in this context, tolerability and compliance will probably become the most important factors in the future, according to the emerging value of quality of life in cancer care. The availability of new drugs, such as Abraxane®, in association with other traditional and nontraditional drugs—new antineoplastic agents and targeted molecules—will give the oncologist many different effective treatment options for patients in this setting."

Researchers also note that a firm working with Abraxis, BIND Biosciences, makes a proprietary Medicinal Nanoengineering™ platform, which is going to play a major role in future nanotech drug design, engineering, and manufacturing of nanoparticle therapeutics with control over drug PK; biodistribution; cell and tissue-specific targeting; as well as drug exposure kinetics. So-called medicinal nanoengineering complements medicinal chemistry to enable predictable and rapid development of breakthrough products with superior efficacy and safety profiles. Initial product development efforts are in the areas of oncology. The firm is working with Abraxane and other drug developers.

Future Prospects for Abraxane

Investors—who CEO Lonnie Moulder cultivated assiduously at the Merrill Lynch health investor conference described above—have showered money upon Abraxis BioScience. Abraxis has been investing the money in expansion.

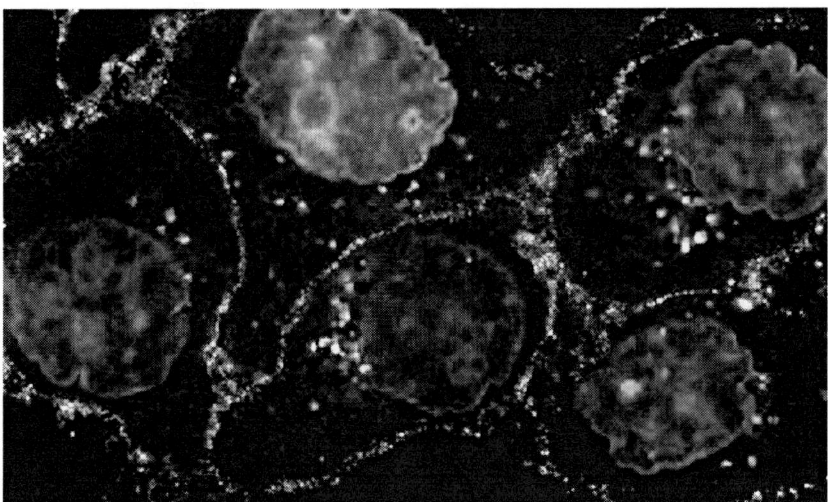

FIGURE 2.1 BIND Biosciences' targeted nanoparticles inside cells. Image courtesy of BIND Biosciences.

Abraxis Health, a division of Abraxis BioScience, Inc., in the summer of 2010, disclosed that it had broken ground for a new manufacturing facility in Phoenix, AZ.

The biopharma manufacturing plant represents a $70 million investment by Los Angeles-based Abraxis and will create up to 200 high-tech jobs. The new facility will produce Abraxane, the solvent-free chemotherapy treatment option for metastatic breast cancer, which was developed using Abraxis BioScience's proprietary *nanoparticle albumin-bound (nab®)* technology platform.

The facility will be the world's largest—and most sophisticated—protein nanobiologics manufacturing plant, with annual production capacity of over 10 million units and the capability to expand to 20 million units, suitable for worldwide distribution, according to company executives.

The factory, formerly owned by Watson Pharmaceuticals, was bought by Abraxis in July 2007, which even then had an eye toward future expansion.

Abraxis has hired more than 100 former Watson employees, recruited additional staff from among the local workforce, and will continue to recruit as the facility ramps up production.

"As we continue to grow, these new premises in Phoenix will become the flagship manufacturing facility for our cutting-edge technology," said founder and Chairman Dr. Patrick Soon-Shiong, who is still with the company, now an operating unit of Celgene.

"This is great news for Phoenix," said Mayor Phil Gordon, adding that with this plant, Abraxis was "bringing highly skilled jobs and a welcome boost for our economy."

The company now has multiple agreements in place for the development of novel, insoluble compounds using the nanobiologics technology employed at the Phoenix plant, the company's founder said.

But even while the new plant was coming online, the doctors at Abraxis were continuing to come up with new uses for their technology.

The company's tumor-targeting technology, the nab platform, and secreted protein acidic and rich in cysteine (SPARC) biomarker is being shown to have even more uses, as the previously cited Italian researchers noted in their landmark paper on cancer and nanotechnology.

One new study, undertaken by the company, not outside researchers, examined the relationship between a tumor biomarker, SPARC, and tumor blood vessel growth and metastasis, also referred to as *angiogenesis*.

The findings, the company said, showed that the role of Abraxis' novel chemotherapy agent Abraxane in increased anti-angiogenic and anti-tumor activity was due to the unique albumin-bound delivery mechanism provided through the nab technology platform.

"Preliminary clinical studies have suggested a direct correlation between elevated levels of protein secretions and positive response to Abraxane," said Patrick Soon-Shiong, MD, chairman and chief executive officer of Abraxis BioScience. "These new data improve our understanding of protein secretions and further confirms the benefits of albumin-bound nab technology in enhancing delivery of chemotherapy."

Simply put, Abraxane's nab technology exploits tumor biology to deliver more effective and targeted chemotherapy treatment. Generally, with normal tumor biology, the proteins secreted by the genes of the tumor would be used to attract nutrients conveyed by naturally occurring albumin.

But the research proved the concept that, with Abraxane's nab technology, these secretions bind to the albumin-bound medicine. Rather than delivering nutrients to the tumor, it provides chemotherapy, killing cancer cells and completely stopping tumor growth.

The study, company researchers said, was designed to define the role of proteins in fostering blood vessel growth and tumor invasiveness. Doctors were also looking for ways to characterize the protein's angiogenic and albumin binding domains—the metaphorical lock and key formations that allow albumin to bind to the protein.

Other findings of note, recently, for researchers at Abraxis include the following:

- Employing recombinant human secreted protein acidic and rich in cysteine (rhSPARC) grown *in vitro*, doctors determined that rhSPARC fostered the growth of new blood vessels (pro-angiogenesis) at physiological levels of SPARC found in cancer tumors.
- By adding rhSPARC to the drug delivery mechanism, researchers found that there was the development of more mature blood vessels well supported by pericytes (smaller, supporting blood vessels wrapped around capillaries). This

implies that SPARC plays a bigger role in the angiogenic process beyond initiating the growth of new blood vessels.

- Doctors determined that the protein's angiogenic domain was located on the carboxyl, or C-terminus, of the SPARC molecule.
- Researchers found that the albumin binding domain was found to be localized to amino acids 209–223 of the SPARC molecule.
- Doctors found that rhSPARC and albumin were binding sustainably at levels similar to the plasma concentration of albumin in the body.

"Abraxis' research on SPARC is part of our ongoing commitment to pursue scientific models that support giving the right treatment to the right patient at the right time," said Neil Desai, PhD, senior vice president, global research and development. "This research further confirms that our nab technology platform provides a novel approach to cancer treatment through a unique delivery mechanism."

Further upstream in the drug delivery pipeline, studies suggest that new drugs ABI-009 (nab-rapamycin) and the vascular disruptive agent ABI-011 (nab-5404). ABI-009 is a signal transduction inhibitor that targets the mTOR pathway, which is completely critical to cell proliferation and survival. The investigational agent ABI-011 is designed to directly disrupt tumor vasculature to compromise tumor growth and progression. Both of these agents utilize Abraxis' proprietary nab delivery mechanism and are currently being studied for a wide range of solid tumor types, the company said.

Rivals Emerging

As one might expect, others are not letting Abraxis keep this emerging market all to itself. The global nanotechnology market is expected to be $27 billion by 2015, according to a new report by BCC Research. The compound annual growth rate of the market is expected to be 11.1 percent, according to the study, "Nanotechnology: A Realistic Market Assessment."

According to these researchers, second-generation nanotherapeutics to treat cancer are leading the pack in terms of potential for growth. These drugs have drawn a lot of attention—and new funding—as companies seek to improve the efficacy of their early-stage treatments and expand their applications, just like Abraxis has done.

Usually in the size range of 1 to 100 nm, these new drugs have emerged as novel antitumor agents because they have the potential to deliver "high concentrations of drugs to cancer cells" and cause less toxicity than systemically administered drugs, according to BCC Research.

The Food and Drug Administration (FDA) has approved for marketing cancer nanodrugs that are made up of chemotherapeutic agents formulated with liposomes that are bound to the protein albumin. The first nanodrug sanctioned by the FDA for breast cancer was Abraxis BioScience's Abraxane.

Another approved nanodrug is pegylated liposomal doxorubicin, which is marketed as Doxil in the U.S.A. by Centocor Ortho

Biotech, and in Israel by Janssen-Cilag, for ovarian cancer that has progressed after platinum-based chemotherapy.

Schering-Plough has exclusive rights to the medication, branded as Caelyx, in the rest of the world, save for Japan, for the same indications as well as AIDS-related Kaposi's sarcoma, another form of cancer. Regulators quickly approved Doxil/Caelyx due to the belief that the reduction of associated, significant toxicities including cardiomyopathy, bone marrow depression, alopecia, and nausea would benefit patients.

Others have joined the fray more recently. Sopherion Therapeutics has created a nonpegylated form of the drug, called Myocet, and is seeking to expand approval into the first-line setting for Her2-positive metastatic breast cancer. The FDA recently granted Fast Track designation for this indication. A Phase 1II comparison of free doxorubicin versus Myocet, used in combination with cyclophosphamide, to treat advanced breast cancer demonstrated that the incidence of cardiac events was 29 percent versus 13 percent and of congestive heart failure was 8 percent versus 2 percent for the doxorubicin group and the Myocet group, respectively, according to BCC Research. Both groups, however, achieved comparable response rates of about 26 percent and "progression-free survival times" of approximately four months, according to researchers.

Pharmaceutical firms are trying to further bolster the value of anticancer nanodrugs beyond reduced toxicity. The goal: create second-generation nanotherapeutics that have greater specificity, target selectivity, and a vastly improved therapeutic index. Preclinical researchers have demonstrated that doxorubicin-loaded liposomes conjugated with folic acid can be internalized when binding with folate receptors on cancer cells. This indicates the potential for improved targeting and a resulting internalization strategy in the treatment of several multidrug-resistant tumors. What is more, doxorubicin-loaded long-circulating liposomes modified with RGD-peptide motif can reportedly target the neovasculature of angiogenic tumors, according to researchers.

Celgene's own cancer products had generated revenues of close to $2.7 billion in 2009, mostly from its multiple myeloma drug, Revlimid.

Abraxis is seeking to expand Abraxane's use into lung and pancreatic cancers. Earlier in June, Abraxis and Specialized Therapeutics Australia reported positive Phase 1II results in advanced non-small-cell lung cancer (NSCLC) when compared with Taxol in the first-line setting. The study was conducted at 102 sites globally and enrolled 1052 patients. It showed that patients treated with Abraxane and carboplatin demonstrated a significant improvement in overall tumor response rate compared with patients treated with Taxol and carboplatin.

Should the FDA approve Abraxane for treatment of NSCLC, Celgene will have to pay Abraxis $250 million. The company will be further obligated to pay $300 million if the drug gains FDA approval for treating pancreatic cancer, and yet another $100 million if this cancer sanction comes by April 1, 2013.

Abraxis says that the delivery specificity occurs through targeting a tumor-activated, albumin-specific biologic pathway with the albumin nanoshell. This nanoshuttle system then activates an albumin-specific (Gp60) receptor-mediated transcytosis path through proliferating tumor cell walls. Once in the stromal microenvironment, the albumin-bound drug may be preferentially localized by a second albumin-specific binding protein, SPARC, which is secreted into the stroma by tumor cells. The resulting collapse of stroma surrounding the tumor cell may thus enhance the delivery of the nab-chemotherapeutic to the intracellular core of the tumor cell itself.

Another firm in the nanodrug development space is BIND Biosciences, mentioned briefly above. In the summer of 2010, it announced that it had raised $12.4 million in a Series C-1 financing. The proceeds are being used to conduct initial clinical trials on its lead drug candidate, BIND-014, for patients with taxane-sensitive tumors such as prostate, breast, and lung cancers.

BIND CEO, Scott Minick, told reporters recently that he had been looking for next-generation therapeutic nanotechnology platforms as a venture capitalist for eight years before joining BIND. "We decided to invest in the BIND technology because it had the properties that could improve drugs, reduce toxicity, increase efficacy, and could work for a wide range of different drugs including small molecules, peptides, proteins, and siRNA," he said in an interview with a trade journal.

"Additionally, it was technology that could be manufactured; nanoparticle technology is challenging to scale. Translating manufacturing from the bench to kilograms has been difficult to impossible for a lot of these technologies, but BIND has already addressed it effectively," he told the trade publication.

Minick distinguished BIND's approach from those of first-generation liposome-based nanoparticles, further noting that the company's "medicinal nanoengineering allows it to optimize properties in defined ways to achieve the biological result we want. We target through two different mechanisms. Biophysical properties of the nanoparticles are engineered to fit through gaps in blood vessels surrounding tumors and other disease sites. We also put a targeting ligand on the surface of the nanoparticle to bind to specific cell surface markers. And a therapeutic payload is incorporated into the targeted nanoparticle," he told the journal.

The company's biocompatible and biodegradable polymer matrix for drug encapsulation has been used in humans many times, Minick said. "The PEG coating is the same PEG coating used in stealth liposomes, and it forms a hydration shell," he told *Genetic Engineering News*. These nanoparticles are apparently the same size as a virus but look more like water droplets to the immune system. Thus, the immune system ignores them. The breakdown products of PLA and PLGA are benign compounds: lactic acid and glycolic acid.

Minick added, "In the case of our first drug, our ligand binds to PMSA, a protein that is expressed on certain tumor cells such as prostate cancer cells as well as tumor neovasculature. The key differentiating property is that we can improve the underlying drug—docetaxel in the case of Bind-014. We have seen in animal models that we get up to a 20-fold increase in drug concentration in the tumor compared to administering the free drug."

Minick also commented that first-generation liposome-based nanotechnologies enabled "far fewer degrees of freedom and far fewer drugs relative to ours. It's quite easy for us to change size, charge, etc."

Experts note that new ideas about how to optimize nanoparticles for drug delivery in cancer continue to emerge. One of the latest approaches involves nanoscale assemblies that may be activated through external stimuli, which developers say may represent the next stage in multifunctional cancer therapeutics.

Intense Interest Overseas

These drugs are not just for domestic consumption—there is an emerging worldwide market for them. On June 29, 2010, New Zealand and Japan announced that they had approved Abraxis' albumin-bound paclitaxel for breast cancer therapy there.

Abraxis BioScience reported regulatory approval of Abraxane® in New Zealand for the treatment of metastatic breast cancer—once physicians there report failure of anthracylcine therapy. The approval for the albumin-bound paclitaxel formulation follows on from Japanese clearance, which was granted just a prior to the New Zealand news.

The product will be distributed in New Zealand by marketing partner Specialized Therapeutics once reimbursement approval has been granted by the country's pharmaceutical reimbursement authority. Meanwhile, in Japan, the drug will be marketed by Otsuka's Taiho Pharmaceutical business.

Abraxis is exploiting its technology for the development of nab-based formulation of docetaxel and rapamycin, which are in Phase II and Phase I development, respectively. A nab-formulated Hsp90 inhibitor and a dual microtubule/topoisomerase-1 inhibitor are also in preclinical development.

Regulators Assuage Public on Safety of Nanotechnology Drugs

For years, there has been a concern among regulators that the public might become alarmed at the approval of nanotech drugs. Back during the 1990s, there were worries over "frankenfoods" when genetic engineering first gained a lot of media attention, even though hybridization has been going on since Augustinian monk Gregor Mendel cross pollinated plants at his monastery in nineteenth century Europe.

As far back as 2006, during the Bush administration, the FDA held public meetings to inform health and public safety activists of the safety of nanotechnology drugs that had been approved or were in the process of being approved.

On Oct. 2, 2006, the FDA held its first public meeting about the "regulation" of nanotech in Bethesda, Md., according to the *American Journal of Public Health.*

"FDA recognizes that nanotechnology has a great potential to promote public health through advances in medical products," said Randall Lutter, co-chair of FDA's 23-member nanotechnology task force, trying to get out in front of the pack and set the PR agenda for discussion of nanotechnology in medicine. This seminar happened just after Abraxane was approved for anticancer therapy.

Lutter added that he hoped the meeting would "increase awareness of both the challenges and the opportunities that nanotechnology may provide and how we can best meet those challenges and opportunities."

A nanometer is a billionth of a meter, less than half the diameter of a DNA strand, Lutter explained to the crowd.

There was some pushback from interest groups, but it was rather tame as compared to earlier efforts to discredit the genetic engineering of foods, and even irradiation of fruits and vegetables, controversial in some circles in the 1990s.

"FDA should undertake the difficult but important step to develop clear definitions and nomenclature for nanoengineered materials and nanotechnology, both for regulatory purposes and for minimizing consumer confusion," said Consumers Union representative Carolyn Cairns, according to the report in the *American Journal of Public Health.*

Particular concern was raised at the meeting as to the safety of nanotechnology cancer therapies. Among the unknowns about the emerging science of nanotechnology was whether nanoscale products would behave in unexpected ways that result in unacceptable risks to humans and the environment.

"We're deeply concerned that the unregulated widespread use of many nanoengineered substances may generate the types of irreversible unintended consequences seen before with other innovative materials, such as polychlorinated biphenyls and pesticides like DDT, that were pushed to market before their risks were characterized," Cairns said, according to the report in the *American Journal of Public Health.* "We encourage FDA to revise its priorities with a greater emphasis on protecting consumers from nanotechnology's adverse effects than on removing hurdles that inhibit its use."

Philippe Martin, of the European Commission, said that European regulators currently view nanotechnology products as unpredictable. "Small is small, small is different, and small is hard to predict," Martin said, according to the report. "A billionth of a meter is also small with respect to the natural barriers to the entry and the movement of particles in the human body."

This uncertainty persists in Europe but does not seem to worry Americans too much. "We cannot assume that what we know about the bulk substance applies to the nano substance," Martin said, according to the journal report. Thus, regulators "have to operate on a case-by-case basis when evaluating nanotechnology products."

At the meeting, it emerged, interestingly, that the FDA has taken the stance that most drugs become nanoscale "at some point during metabolism" and absorb in the body, according to the report. Since the FDA knows of no particle-size-related safety concerns that have applied to drugs, the agency has argued that the existing regulatory framework is good enough for regulating nanotechnology drugs.

The FDA seems to have achieved its strategic communications goal with the 2006 meeting. There have been no mass-scale protests, letter-writing campaigns, or bad PR events for regulators, or the developers, to contend with from activists.

Futuristic Vision

According to a report in *Nature* magazine, if nanotechnology is fully integrated with the established cancer research enterprise, nanotechnology might revolutionize cancer treatment and, more than 40 years after the war on cancer was declared by President Nixon, actually bring that war to an end with a heroic victory.

Some of the principal challenges to compete victory over cancer, however, remain. These include the following.

Developing approaches for the in vivo detection and monitoring of cancer markers

The early detection of precancerous and neoplastic lesions remains an elusive goal. Cancer imaging technologies do not possess sufficient spatial resolution for early detection based on lesion anatomy. To pinpoint malignancies based on their molecular profiles, all imaging technologies require contrast agents, comprising a signal-amplifying material linked to a molecular recognition and targeting agent, like an antibody.

Nanoparticle technologies are under development as candidates for multifunctional, molecularly or physically targeted contrast agents for every clinical imaging modality, with the goal of detecting earlier-stage cancer tumors, finding molecular expressions of neoplasms and their microenvironment, and providing improved anatomical definition for cancerous lesions.

For example, researchers recently demonstrated that lymphotropic paramagnetic nanoparticles enable the MRI imaging of clinically occult lymph-node metastases in patients with prostate cancer, which are detectable only by invasive approaches, i.e., surgery or biopsy... Researchers are using polymeric dendrimers as gadolinium nanocarriers to image the lymphatic drainage of breast cancer in rodents, indicating that this procedure could be used clinically instead of sentinel lymph node biopsy.

One research project demonstrated that dextran-coated, ultra-small paramagnetic iron-oxide nanoparticles outperformed conventional gadolinium MRI contrast in terms of intraoperative permanence of imaging enhancement, inflammatory targeting, and detectability at low magnet strength in the surgical treatment of brain tumors. Bimodal nanoparticles, conveying a near-infrared optically detectable fluorochrome linked to an MRI contrast agent—cross-linked iron oxide—were used by doctors for the preoperative, contour-defining imaging of a brain tumor, and the intraoperative visualization.

So-called nanoparticle probes which contain molecularly targeted recognition agents might also provide information as to the presence and distribution of cancer signatures and markers linked with the tumor microenvironment. Iron oxide nanoparticles were conjugated to annexin-V, which finds the phosphatidylserine that is present on apoptotic cells. This was also used for MRI identification of camptothecin-induced apoptosis of Jurkat T cells.

Telomerase activity, a marker of limitless replicative potential, was also found by MRI in cell assays by doctors using biologically 'smart' nanoparticles that switch their magnetic state.

Nanotechnologies for molecular cancer detection, identification, and diagnostics

Microarrays can be used to identify patterns of molecules on surfaces, with sophisticated control over their spatial placement, for example, to obtain DNA sequencing by hybridization on a microchip. The techniques of photolithography have been adapted from the microelectronic industry. Controlling the lateral dimensions of each square in the checkerboard of a microarray was initially on the order of 100 μm, or 100,000 nm. The linear spatial resolution of lithography is 1,000 times better, however, demonstrating that up to a million-fold increase in information density could be packed in these "nanoarrays."

Photolithography techniques can be used to pattern different chemistries, biological moieties, and even physical textures on substrates, to prefractionate protein mixtures before investigation by mass spectrometry. Distinct proteomic patterns are produced by different substrate treatments with each biological sample.

Refining technology platforms for ex vivo, early detection of cancer biomarkers

Serum markers for early detection of most cancers are not yet available. The markers that are in clinical use, such as prostate-specific antigen (PSA) and carcinoembryonic antigen (CEA), are nonspecific, according to physician researchers, have vastly different baseline expressions, and are of limited effectiveness for early detection of cancer.

The goal of developing reliable, nanotechnology-based early-detection approaches from serum, and other fluids, is of paramount importance to medicine, according to *Nature* magazine.

A number of nanotechnologies are sound candidates for early detection platforms, starting with surface-patterning approaches, mentioned above, and including already established technologies such as DNA microarrays and SELDI-TOF mass spectroscopy for proteomics.

The transition from micron-scale to nanoscale dimensional control on surface features would amount to massive increases in information quality and quantity, ushering in entirely new approaches to molecular recognition.

According to *Nature*, researcher James Gimzewski and colleagues recently pioneered the concept of biomolecular binding events. These events yield forces and deformations that might be detected and recognized by selective sensing nanostructures.

Examples of such devices are micro- or nanocantilevers, which change frequencies as a result of affinity binding. Other researchers, led by Dr. Arun Majumdar, used microcantilevers to detect single nucleotide polymorphisms (SNPs) in a 10-mer DNA target oligonucleotide. And that was without employing extrinsic fluorescent or radioactive labeling. These doctors also demonstrated the applicability of microcantilevers for the quantitation of PSA. The sensitivities of these primitive nanotechnology assays do not yet offer true advantages over conventional detection methods, although they are quite promising.

In vivo molecular sensors

A different approach to molecular detection *in vivo* involves the use of implantable sensors, equipped with technology to relay sensed information extra corporeally. Despite many years of research toward this vision, the unsolved challenge for the clinical deployment of implantable molecular sensors remains the unwanted, nonspecific adsorption of serum proteins on the sensing surfaces. This phenomenon is known as *biofouling*, and it results in the rapid loss of the ability of the sensor to detect the protein of interest over the background signal.

According to the National Cancer Institute, an institute of the National Institutes of Health, there are other new detection technologies being developed in labs as of late October 2010. The NCI is funding this early-stage research. According to a feature on the NCI's web site (accessed October 25, 2010, at *http://nano.cancer.gov/action/news/2010/oct/nanotech_news_2010–10–25a.asp*), and entitled, *Tracking Tumor Targeting Nanoparticles in the Body,* researchers still don't have a solid understanding of how nanoparticles reach tumors and how they bind to and enter the targeted tumor.

Two teams of investigators, both part of the Alliance for Nanotechnology in Cancer, have undertaken studies aiming to track nanoparticles as they move through animals.

One team of investigators at Stanford University used quantum dots to study how nanoparticles travel through tumor blood vessels in living animal subjects, connect to molecular targets on the surface of the tumor, and then travel into the tumor itself. Sanjiv Sam Gambhir, co-director of one of the nine National Cancer Institute (NCI) Centers of Cancer Nanotechnology Excellence, led this project. Gambhir and his associates published their research in the journal *Small.* In another study, published in the journal *ACS Nano,* NCI researchers Dong Shin, Mostafa El-Sayed, and Shuming Nie of Emory University and the Georgia Institute of Technology used targeted gold nanocrystals to examine active and passive tumor targeting.

Dr. Gambhir, of Stanford, and his collaborators utilized the capabilities of intravital microscopy, a technique that enables doctors to see fluorescent markers through a live animal's skin. The Stanford team eyed nanoparticles in mice in which a number of types of tumors were permitted to grow in the ears of the animals. The investigators used a molecule known by medicine to bind tightly to a protein found on the surface of tumor blood vessels.

Researchers found that whatever the type of tumor studied, nanoparticle binding occurred only when aggregates of particles tethered themselves to multiple sites within a tumor. The doctors weren't able to find any significant binding when they repeated these experiments. The researchers also determined that binding rates and patterns were consistent for all tumor types, which the NCI considers "reassuring," given the natural heterogeneity that characterizes cancers.

Meantime, the Emory-Georgia Tech team deployed rod-shaped gold nanocrystals connected to tumor-targeting peptides to examine the delivery mechanisms that allow nanoparticles to grow in tumors. Doctors used gold nanoparticles so that they could determine how many nanoparticles reached tumors. Gold doesn't occur naturally in mammals; thus, gold detected in a given tumor or tissue using the technique known as *elemental mass spectrometry* would come from gold nanoparticles.

Doctors created three formulations, attaching one of three tumor-targeting molecules to the surface of gold nanorods. The researchers injected the nanoparticles into animals with implanted human tumors. Doctors then allowed the nanoparticles to circulate, after which they measured the amount of gold that accumulated in the implanted tumors and other tissues. The doctors also conducted this experiment another time, using untargeted gold nanoparticles. Targeting molecules only marginally increased the amount of gold that accumulated in tumors.

The investigators concluded that gold nanoparticles designed to be used in photothermal anticancer therapy should be injected directly into tumors rather than via intravenous administration to achieve the greatest concentration of gold in tumors. They also noted in their paper that these experiments suggest that target binding is not the rate-limiting step for nanoparticle delivery but, rather, that transport out of the blood stream and into tumors is the major barrier to nanoparticle accumulation in tumors.

The work involving intra-vital microscopy, which is detailed in a paper titled, "Dynamic Visualization of RGD-Quantum Dot Binding to Tumor Neovasculature and Extravasation in Multiple Living Mouse Models Using Intravital Microscopy," was supported by the *NCI Alliance for Nanotechnology in Cancer*, a comprehensive initiative designed to accelerate the application of nanotechnology to the prevention, diagnosis, and treatment of cancer.

Voilá! Moment for Researchers Finally Arrives

Famed publisher Steve Forbes often uses the line, "with all thy getting, get understanding," in his biweekly column in

Forbes magazine. That somewhat archaic phrase* is remarkably apt for the futuristic world of cancer nanotechnology.

A few years after Abraxis BioScience first obtained FDA approval for its first ever anticancer nanotechnology drug, medicine is finally actually starting to understand how nanoparticles interact with cells.

According to a feature on the web site of the National Cancer Institute,[2] medical researcher Dr. Chad Mirkin, co-principal investigator at the Northwestern University Center for Cancer Nanotechnology Excellence (one of nine such centers created by the National Cancer Institute [NCI]), studied how cancer cells respond to different nucleic acid-nanoparticle formulations. Mirkin and his team of investigators published their research in the journal *ACS Nano.*

To determine just how cancer cells respond when they take up nanoparticles, Mirkin and his team relied upon a technique called *genome-wide expression profiling*. This technique measures relative changes in gene expression.

Research investigators inserted different kinds of nanoparticles into cancer cells that were growing in culture dishes. Then they obtained entire genome expression profile for each of the cells. Researchers affixed non-targeting nucleic acids to the nanoparticles to regulate the changes in the cells.

As a result of these studies, the doctors showed that surface properties of the nanoparticles significantly affected how a particular nanoparticle impacted gene expression inside the cell.

---

* Orig. from Proverbs 4:7.

Researchers compared two nanoparticles that differed in the way in which their respective nucleic acids were attached to the nanoparticle surface.

Linking nanoparticles to nucleic acids through a covalent chemical bond had almost no effect on gene expression, the doctors found. This research, the doctors noted, demonstrated just how important it is to completely characterize nanoparticles not just in terms of the shape and size, but also with regard to their unique "surface properties," according to the NCI.

References

[1] Miele, Evelina, Gian Paolo Spinelli, Federica Tomao, and Silverio Tomao. Albumin-Bound Formulation of Paclitaxel (Abraxane® ABI-007) in the Treatment of Breast Cancer. *Int J Nanomedicine* 2009, 4:99–105. PMCID: PMC2720743, published online April 20, 2009.

[2] Mirkin, Chad. Understanding How Cells Respond to Nanoparticles, http://nano.cancer.gov/action/news/2010/oct/nanotech_news_2010–10–25c.asp, accessed Oct. 25, 2010.

Suggested Reading

Jemal, A., R. Siegel, E. Ward, et al. Cancer Statistics, 2008. *CA Cancer J Clin* 2008; 58(2):71–96. [*PubMed*]

Newman, L.A., and S. E. Singletary. Overview of Adjuvant Systemic Therapy in Early Stage Breast Cancer. *Surg Clin North Am* 2007; 87(2):499–509. [*PubMed*]

Guarneri V., and P. F. Conte. The Curability of Breast Cancer and the Treatment of Advanced Disease. *Eur J Nucl Med Mol Imaging* 2004; 31 (suppl. 1):S149–S161. [*PubMed*]

Bristol-Myers Squibb Co. (Princeton, NJ) 2003. Taxol® (paclitaxel) injection, package insert.

Aventis Pharmaceutical Products, Inc. (Bridgewater, NJ); 2003. Taxotere® (docetaxel) injection concentrate, package insert.

National Comprehensive Cancer Network, Clinical Practice Guidelines in Oncology: Breast Cancer v2 2008. Available *http://www.nccn.org/professionals/physician_gls/default.asp*.

Schiff, P. B., and S. B. Horwitz. Taxol Stabilizes Microtubules in Mouse Fibroblast Cells. *Proc Natl Acad Sci USA* 1980; 77:1561–1565. [*PMC free article*] [*PubMed*]

Rowinsky, E. K., L. A. Cazenave, and R. C. Donehower. Taxol: a Novel Investigational Antimicrotubule Agent. *J Natl Cancer Inst* 1990; 82:1247–1259. [*PubMed*]

Schiff, P. B., J. Fant, and S. B. Horwitz. Promotion of Microtubule Assembly *in vitro* by Taxol. *Nature* 1979; 277:665–667. [*PubMed*]

Rowinsky, E. K., and R. C. Donehower. Paclitaxel (Taxol) *N Engl J Med* 1995; 332:1004–1014. [*PubMed*]

Chu, E., and V. T. DeVita, 2002, 2003. *Physicians' Cancer Chemotherapy Drug Manual*. Sudbury, MA: Jones and Bartlett Publishers, p. 284.

Choy, H. Taxanes in Combined-Modality Therapy for Solid Tumors. *Oncology* 1999; 13:23–38. [*PubMed*]

Hainsworth, J. D. Practical Aspects of Weekly Docetaxel Administration Schedules. *Oncologist* 2004; 9:538–545. [*PubMed*]

Crown, J., and M. O'Leary. The Taxanes: An Update. *Lancet* 2000; 355:1176–1178. [*PubMed*]

Wani, M. C., H. L. Taylor, M. E. Wall, P. Coggon, and A. T. McPhail. The Isolation and Structure of Taxol, a Novel Antileukemic and Antitumor Agent from Taxus brevifolia. *J Am Chem Soc* 1971; 93:2325–2327. [*PubMed*]

Adams, J. D., K. P. Flora, B. R. Goldspiel, J. W. Wilson, S. G. Arbuck, and R. Finley. Taxol: A History of Pharmaceutical Development and Current Pharmaceutical Concerns. *J Natl Cancer Inst Monogr* 1993; 15:141–147. [*PubMed*]

Riondel, J., M. Jacrot, F. Picot, H. Beriel, C. Mouriquand, and P. Potier. Therapeutic Response to Taxol of Six Human Tumors Xenografted into Nude Mice. *Cancer Chemother Pharmacol* 1986; 17:137–142. [*PubMed*]

Spencer, C. M., and D. Faulds. Paclitaxel: A Review of Its Pharmacodynamic and Pharmacokinetic Properties and Therapeutic Potential in the Treatment of Cancer. *Drugs* 1994; 48:794–847. [*PubMed*]

Onetto, N., R. Canett, B. Winograd, et al. Overview of Paclitaxel Safety. *J Natl Cancer Inst Monogr* 1993; 15:131–139. [*PubMed*]

Bissery, M. C. Preclinical Pharmacology of Docetaxel. *Eur J Cancer* 1995; 31A (suppl. 4):S1–S6. [*PubMed*]

Kloover, J. S., M. A. den Bakker, H. Gelderblom, et al. Fatal Outcome of a Hypersensitivity Reaction to Paclitaxel: A Critical Review of Premedication Regimens. *Br J Cancer* 2004; 90:304–305. [*PMC free article*] [*PubMed*]

Tije, A. J., J. Verweij, W. J. Loos, et al. Pharmacological Effects of Formulation Vehicles: Implications for Cancer Chemotherapy. *Clin Pharmacokinet* 2003; 42:665–685. [*PubMed*]

Sparreboom, A., C. D. Scripture, V. Trieu, et al. Comparative Preclinical and Clinical Pharmacokinetics of a Cremophor-free, Nanoparticle Albumin-Bound Paclitaxel (ABI-007) and Paclitaxel Formulated in Cremophor (Taxol). *Clin Cancer Res* 2005; 11:4136–4243. [*PubMed*]

Winer, E., D. Berry, D. Duggan, et al. Failure of Higher-Dose Paclitaxel to Improve Outcome in Patients with Metastatic Breast Cancer: Cancer and Leukemia Group B Trial 9342. *J Clin Oncol* 2004; 22:2061–2068. [*PubMed*]

Gelderblom, H., J. Verweij, K. Nooter, et al. Cremophor EL: The Drawbacks and Advantages of Vehicle Selection for Drug Formulation. *Eur J Cancer* 2001; 37:1590–1598. [*PubMed*]

Weiss, R. B., R. C. Donehower, P. H. Wiernik, et al. Hypersensitivity Reactions from Taxol. *J Clin Oncol* 1990; 8:1263–1268. [*PubMed*]

Nyman, D. W., K. J. Campbell, E. Hersh, et al. Phase I and Pharmacokinetics Trial of ABI-007, a Novel Nanoparticle Formulation of Paclitaxel in Patients with Advanced Nonhematologic Malignancies. *J Clin Oncol* 2005; 23(31):7785–7793. [*PubMed*]

Sparreboom, A., L. van Zuylen, E. Brouwer, et al. Cremophor Elmediated Alteration of Paclitaxel Distribution in Human Blood: Clinical Pharmacokinetic Implications. *Cancer Res* 1999; 59:1454–1457. [*PubMed*]

Lorenz, W., H. J. Reimann, A. Schmal, et al. Histamine Release in Dogs by Cremophor EL and Its Derivatives: Oxethylated Oleic Acid is the Most Effective Constituent. *Agents Actions* 1977; 7:63–67. [*PubMed*]

Gianni, L., C. M. Kearns, A. Giani, et al. Nonlinear Pharmacokinetics and Metabolism of Paclitaxel and its Pharmacokinetic/Pharmacodynamic Relationships in Humans. *J Clin Oncol* 1995; 13:180–190. [*PubMed*]

Vaishampayan, U., L. van Zuylen, J. Verweij, and A. Sparreboom. Role of Formulation Vehicles in Taxane Pharmacology. *Invest New Drugs* 2001; 19:125–141. [*PubMed*]

Piccart, M. J., J. Klijn, R. Paridaens, et al. Steroids Do Reduce the Severity and Delay the Onset of Docetaxel (DXT) Induced Fluid Retention: Final Results of a Randomized Trial of the EORTC Investigational Drug Branch for Breast Cancer (IDBBC). *Eur J Cancer* 1995; 31A (suppl. 5):S75.

Piccart, M. J., J. Klijn, R. Paridaens, et al. Corticosteroids Significantly Delay the Onset of Docetaxel-Induced Fluid Retention: Final Results of a Randomized Study of the European Organization for Research and Treatment of Cancer Investigational Drug Branch for Breast Cancer. *J Clin Oncol* 1997; 15:3149–3155. [*PubMed*]

Thei, T., D. Peter, K. E. Drexler, et al. Nanotechnology. *Nat Nanotechnol* 2006; 1:8–10. [*PubMed*]

Cloninger, M. J. Biological Applications of Dendrimers. *Curr Opin Chem Biol* 2002; 6:742–748. [*PubMed*]

Rivera, E. Liposomal Anthracyclines in Metastatic Breast Cancer: Clinical Update. *Oncologist* 2003; 8(suppl. 2):3–9. [*PubMed*]

Pan, B., D. Cui, Y. Sheng, C. Ozkan, F. Gao, R. He, et al. Dendrimer-Modified Magnetic Nanoparticles Enhance Efficiency of Gene Delivery System. *Cancer Res* 2007; 67:8156–8163. [*PubMed*]

Oyewumi, M. O., and R. J. Mumper. Engineering Tumor-Targeted Gadolinium Hexanedione Nanoparticles for Potential Application in Neutron Capture Therapy. *Bioconjug Chem* 2002; 13:1328–1335. [*PubMed*]

Yan, F., H. Xu, J. Anker, R. Kopelman, B. Ross, A. Rehemtulla, et al. Synthesis and Characterization of Silica-Embedded Iron Oxide Nanoparticles for Magnetic Resonance Imaging. *J Nanosci Nanotechnol* 2004; 4:72–76. [*PubMed*]

Duncan, R. The Dawning Era of Polymer Therapeutics. *Nat Rev Drug Discov* 2003; 2:347–360. [*PubMed*]

Green, J. J., E. Chiu, E. S. Leshchiner, J. Shi, R. Langer, AND D. G. Anderson. Electrostatic Ligand Coatings of Nanoparticles Enable Ligand-Specific Gene Delivery to Human Primary Cells. *Nano Lett* 2007; 7:874–879. [*PubMed*]

Hirsch, L. R., R. J. Stafford, J. A. Bankson, S.R. Sershen, B. Rivera, R. E. Price, et al. Nanoshell-Mediated Near-Infrared Thermal Therapy of Tumors under Magnetic Resonance Guidance. *Proc Natl Acad Sci USA* 2003; 100:13549–13554. [*PMC free article*] [*PubMed*]

Loo, C., A. Lowery, N. Halas, J. West, and R. Drezek. Immunotargeted Nanoshells for Integrated Cancer Imaging and Therapy. *Nano Lett* 2005; 5:709–711. [*PubMed*]

Yan, F., and R. Kopelman. The Embedding of Meta-tetra (hydroxyphenyl)-chlorin into Silica Nanoparticle Platforms for Photodynamic Therapy and Their Singlet Oxygen Production and pH-dependent Optical Properties. *Photochem Photobiol* 2003; 78:587–591. [*PubMed*]

Martin, F. J., K. Melnik, T. West, J. Shapiro, M. Cohen, A. A. Boiarski, et al. Acute Toxicity of Intravenously Administered Microfabricated Silicon Dioxide Drug Delivery Particles in Mice: Preliminary Findings. *Drugs R D* 2005; 6:71–81. [*PubMed*]

Peng, J., X. He, K. Wang, W. Tan, H. Li, X. Xing, et al. An Antisense Oligonucleotide Carrier Based on Amino Silica Nanoparticles for Antisense Inhibition of Cancer Cells. *Nanomedicine* 2006; 2:113–120. [*PubMed*]

Kam, N. W., M. O'Connell, J. A. Wisdom, and H. Dai. Carbon Nanotubes as Multifunctional Biological Transporters and Near-Infrared Agents for Selective Cancer Cell Destruction. *Proc Natl Acad Sci USA* 2005; 102:11600–11605. [*PMC free article*] [*PubMed*]

Yong, K. T., J. Qian, I. Roy, H. H. Lee, E. J. Bergey, K. M. Tramposch, et al. Quantum Rod Bioconjugates as Targeted Probes for Confocal and Two-Photon Fluorescence Imaging of Cancer Cells. *Nano Lett* 2007; 7:761–765. [*PubMed*]

Ferrari, M. Cancer Nanotechnology: Opportunities and Challenges. *Nat Rev Cancer* 2005; 5:161–171. [*PubMed*]

Di Paolo, A. Liposomal Anticancer Therapy: Pharmacokinetic and Clinical Aspects. *J Chemother* 2004; 16(suppl. 4):90–93. [*PubMed*]

Maeda, H. The Enhanced Permeability and Retention (EPR) Effect in Tumor Vasculature: The Key Role of Tumor-Selective Macromolecular Drug Targeting. *Adv Enzyme Regul* 2001; 41:189–207. [*PubMed*]

Maeda, H, J. Fang, T. Inutsuka, and Y. Kitamoto. Vascular Permeability Enhancement in Solid Tumor: Various Factors, Mechanisms Involved and Its Implications. *Int Immunopharmacol* 2003; 3:319–328. [*PubMed*]

Gabizon, A., H. Shmeeda, A. T. Horowitz, and S. Zalipsky. Tumor Cell Targeting of Liposome-Entrapped Drugs with Phospholipid-Anchored Folic Acid-PEG Conjugates. *Adv Drug Deliv Rev* 2004; 56:1177–1192. [*PubMed*]

Farokhzad, O. C., J. Cheng, B. A. Teply, I. Sherifi, S. Jon, P. W. Kantoff, et al. Targeted Nanoparticle-Aptamer Bioconjugates for Cancer Chemotherapy *in vivo*. *Proc Natl Acad Sci USA* 2006; 103:315–6320.

Tasciotti, E., X. W. Liu, R. Bhavane, K. Plant, A. D. Leonard, B. K. Price, et al. Mesoporous Silicon Particles as a Multistage Delivery System for Imaging and Therapeutic Applications. *Nat Nanotechnol* 2008; 3:151–157. [*PubMed*]

Tanaka, T., P. Decuzzi, M. Cristofanilli, et al. Nanotechnology for Breast Cancer Therapy. *Biomed Microdevices* 2008. DOI 10.1007/s10544–008–9209–0.

Hawkins, M. J., P. Soon-Shiong, and N. Desai. Protein Nanoparticles as Drug Carriers in Clinical Medicine. *Adv Drug Deliv Rev* 2008; 60:876–885. [*PubMed*]

Purcell, M., J. F. Neault, and H. A. Tajmir-Riahi. Interaction of Taxol with Human Serum Albumin. *Biochem Biophys Acta* 2000; 1478:61–68. [*PubMed*]

Paal, K., J. Muller, and L. Hegedus. High Affinity Binding of Paclitaxel to Human Serum Albumin. *Eur J Biochem* 2001; 268:2187–2191. [*PubMed*]

John, T. A., S. M. Vogel, C. Tiruppathi, et al. Quantitative Analysis of Albumin Uptake and Transport in the Rat Microvessel Endothelial Monolayer. *Am J Physiol Lung Cell Mol Physiol* 2003; 284:L187–L196. [*PubMed*]

Minshall, R. D., W. C. Sessa, R. V. Stan, et al. Caveolin Regulation of Endothelial Function. *Am J Physiol Lung Cell Mol Physiol* 2003; 285:L1179–L1183. [*PubMed*]

Vogel, S. M., R. D. Minshall, M. Pilipovic, C. Tiruppathi, and A. B. Malik. Albumin Uptake and Transcytosis in Endothelial Cells *in vivo* Induced by Albumin-Binding Protein. *Am J Physiol Lung Cell Mol Physiol* 2001; 281:L1512–L1522. [*PubMed*]

Tiruppathi, C., W. Song, M. Bergenfeldt, et al. Gp60 Activation Mediates Albumin Transcytosis in Endothelial Cells by Tyrosine Kinase-Dependent Pathway. *J Biol Chem* 1997; 272:25968–25975. [*PubMed*]

Vorum, H. Reversible Ligand Binding to Human Serum Albumin: Theoretical and Clinical Aspects. *Dan Med Bull* 1999; 46:379–399. [*PubMed*]

Schaumburg, I. L., Abraxis Oncology, a Division of American Pharmaceutical Partners, Inc., 2005. Abraxane® prescribing information.

Desai, N., V. Trieu, Z. Yao, et al. Increased Antitumor Activity, Intratumor Paclitaxel Concentrations, and Endothelial Cell Transport of Cremophor-Free, Albumin-Bound Paclitaxel,

ABI-007, Compared with Cremophor-Based Paclitaxel. *Clin Cancer Res* 2006; 12:1317–1324. [*PubMed*]

Desai, N., Z. Yao, V. Trieu, et al. Evidence of a Novel Transporter Mechanism for a Cremophor-free, Protein-Engineered Paclitaxel (ABI-007) and Enhanced *in vivo* Antitumor Activity in an MX-1 Human Breast Tumor Xenograft Model. 25th Annual San Antonio Breast Conference Symposium, San Antonio, TX. December 11–14, 2002.

Ibrahim, N. K., N. Desai, S. Legha, et al. Phase I and Pharmacokinetic Study of ABI-007, a Cremophor-Free, Protein-Stabilized, Nanoparticle Formulation of Paclitaxel. *Clin Cancer Res* 2002; 8:1038–1044. [*PubMed*]

Ibrahim, N. K., B. Samuels, R. Page, et al. Multicenter Phase II Trial of ABI-007, an Albumin-Bound Paclitaxel, in Women with Metastatic Breast Cancer. *J Clin Oncol* 2005; 23(25):6019–6026. [*PubMed*]

Gradishar, W. J., S. Tjulandin, N. Davidson, H. Shaw, N. Desai, P. Bhar, et al. Phase III Trial of Nanoparticle Albumin-Bound Paclitaxel Compared with Polyethylated Castor Oil-Based Paclitaxel in Women with Breast Cancer. *J Clin Oncol* 2005; 23(31):7794–7803. [*PubMed*]

Nyman, D. W., K. J. Campbell, E.H. Kristen Long, et al. Phase I and Pharmacokinetics Trial of ABI-007, a Novel Nanoparticle Formulation of Paclitaxel in Patients with Advanced Nonhematologic Malignancies. *J Clin Oncol* 2005; 23(31):7785–7793. [*PubMed*]

Henderson, I. C., D.A. Berry, G. D. Demetri, et al. Improved Outcomes from Adding Sequential Paclitaxel but not from Escalating Doxorubicin Dose in an Adjuvant Chemotherapy Regimen for Patients with Node-Positive Primary Breast Cancer. *J Clin Oncol* 2003; 21(6):976–83. [*PubMed*]

Wiedenmann, N., D. Valdecanas, N. Hunter, et al. 130-nm Albumin-Bound Paclitaxel Enhances Tumor Radiocurability and Therapeutic Gain. *Clin Cancer Res* 2007;13(6):1868–1874. [*PubMed*]

3

Controlling Cholesterol

Losing weight effortlessly and controlling cholesterol without changing one's diet have been dreams of diet gurus, and portly couch potatoes, for generations it seems. Now the vision is no longer an illusion.

A Northwestern University start-up in 2009 received a $2.5 million investment to develop nanotechnology that relies on gold nanoparticles to treat cardiovascular disease.

The firm, AuraSense LLC, founded by Northwestern University nanotechnology expert Prof. Chad Mirkin, has developed a nanotechnology device that employs nanoparticles to flush cholesterol out of the patient's bloodstream, dramatically reducing the chances of a stroke or heart attack.

Doctors are using gold and other metals for artificial particles so they show up in medical imaging devices. These artificial particles are created to behave like high-density lipoproteins, also commonly known as "good cholesterol," which protect the body from heart disease.

Touted by Northwestern as its "most cited chemist," Mirkin is a member of President Barack Obama's Council of Advisors for

Science and Technology. Back in June 2009, Mirkin was awarded the $500,000 Lemelson-MIT prize, which recognizes scientific innovation.

Mirkin is also known for creating the nanoparticle-based medical diagnostic assays that underpin the Verigene system, as well as Dip-Pen Nanolithography, an ultrahigh-resolution molecule-based printing technique used by other nanotech researchers.

Prior to the investment, AuraSense had not previously reported any fund raising, or private equity investment, to the Securities and Exchange Commission (SEC).

Mirkin and a colleague, Dr. C. Shad Thaxton, of the Mt. Sinai School of Medicine, Northwestern University, is also working on another project: creating synthetic versions of the good cholesterol molecules using the aforementioned gold nanoparticles to retrofit the fatty core generally found in the HDL molecule. The new molecule has been utilized in a pilot study and shown to be effective in removing plaque buildups from blood vessels by absorbing the cholesterol, just as naturally occurring molecules would.

The research at Mt. Sinai has focused on developing nanoparticles for use in different types of imaging; gold nanocrystals were most effective when used in computer tomography, while iron oxide nanocrystals were more effective with MRIs. The goal of the research is to improve the medical imaging of hardening of the arteries due to the buildup of plaques.

The potential for using the nanoparticles as treatment for cardiovascular disease, according to Dr. William O'Neill, executive dean for clinical affairs at the Miller School of Medicine at the University of Miami, "could revolutionize cardiology."

Yet, as Dr. Andre Nel, director of the Center for Environmental Implications of Nanotechnology at UCLA and Prof. Vincent M. Rotello, at the University of Massachusetts at Amherst, note, further research on the likely effects of nanoparticle accumulation in the body need to be done before use of the nano-HDL in treatment for cardiovascular disease can be approved by the Food and Drug Administration (FDA).

Still Early Stage

Research on nanotechnology for cholesterol control is still very early stage, but developments like Mirkin's, cited above, are quite promising. What, exactly, is cholesterol, and what does it have to do with heart disease?

An array of international studies, over the years, have demonstrated that high levels of cholesterol in the blood dramatically increase the risk of developing coronary disease. Simply put, cholesterol is a fat-like, waxy substance that actually occurs naturally in cells throughout the body. The body relies on cholesterol to create cell membranes, as well as hormones and even Vitamin D and bile acids.

However, only a small amount of cholesterol is needed by the body for these purposes. Some scientists speculate that the body can make all the cholesterol it needs without anything extra supplied by the diet.

Researchers note that cholesterol circulates in the blood system and is attached to proteins called *lipoproteins*. The most notable of these proteins are called *low-density lipoproteins (LDLs)* as well as *high-density lipoproteins (HDLs)*. Most of the cholesterol in the blood stream is carried by LDLs, and this kind of cholesterol is commonly referred to as “bad cholesterol.” Excess cholesterol, unused by the body for the above-cited membrane and hormone creation, becomes deposited on the walls of the arteries. This can lead to a narrowing of the arteries, and even a blockage of the arteries, and thus heart attacks and strokes.

Meanwhile, HDL cholesterol is commonly called “good cholesterol.” HDL moves cholesterol from all over the body to the liver, whence it is broken down and disposed of by the body. Studies showing that lowering bad cholesterol levels reduces coronary heart disease. What is more, other studies have shown that lowering cholesterol reduces risk of heart attack or stroke for those who already have been diagnosed with coronary artery disease.

Blood cholesterol is affected by what you eat, and also how fast your body makes bad cholesterol and gets rid of it through normal biological processes. With age, cholesterol levels increase. After menopause, women often have higher cholesterol levels. Generally, doctors recommend a low-tech solution to lowering cholesterol levels—eating fewer foods that contain saturated fats. That’s food that comes mostly from animal meat. Saturated fats raise LDL levels more than any other food factor, doctors note.

Reducing cholesterol is not the only goal of scientists working with nanotechnology. Monitoring cholesterol is as well.

According to an article titled, *Monitoring Blood Cholesterol with Carbon Nanotubes,*[1] research scientists at the at Miami's Florida International University created a new type of working electrode consisting of vertically aligned multiwall carbon nanotubes, all fitting on a single silicon platform. The silicon-based working electrode can be integrated with signal processing circuitry and can be part of a lab-on-a-chip system.

This innovative nanotechnology facilitates the efficient attachment of biomolecules. The fabricated working electrodes shown in Fig. 3.1 demonstrate a linear relationship between

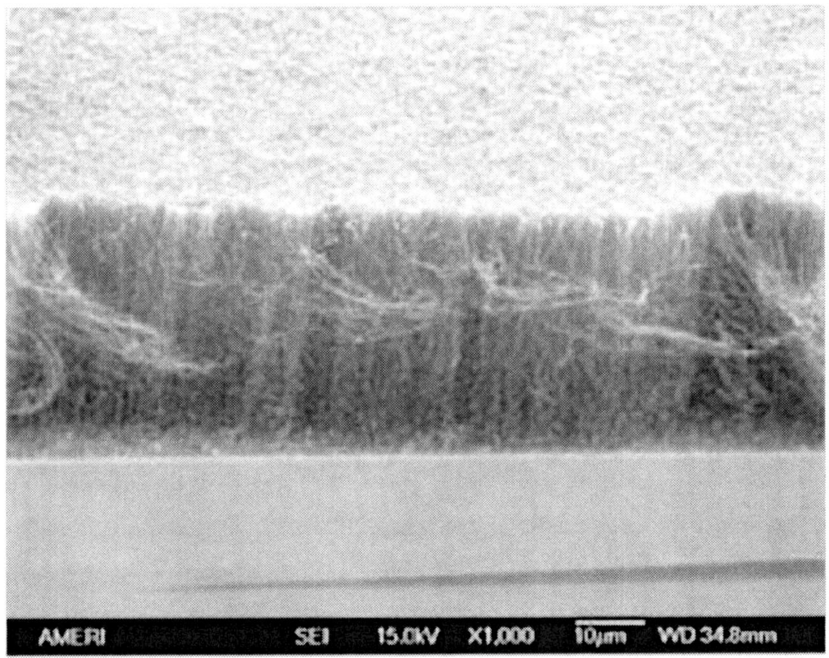

FIGURE 3.1 Nanotubes for cholesterol monitoring. Courtesy of Florida International University.

cholesterol concentration and the output signal. The efficiency of the multiwall carbon nanotubes in facilitating heterogeneous electron transfer was evident by distinct electrochemical peaks and higher signal-to-noise ratio as contrasted with the Au electrode with completely identical enzyme immobilization protocol.

Another interesting use of nanotechnology in cholesterol control is in the creation of designer RNA that can run interference with the metabolism of cholesterol, according to researchers.

Tiny, custom-designed bits of ribonucleic acid (RNA) interfere with cholesterol metabolism, cutting harmful cholesterol in the blood stream by two thirds in preclinical tests, according to a new study by researchers at UT Southwestern Medical Center in collaboration with Alnylam Pharmaceuticals and the Massachusetts Institute of Technology (MIT).

In a report that appeared in the *Proceedings of the National Academy of Sciences,* researchers found that a single dose of a small interfering RNA (siRNA), the chemical cousin of DNA, lowered cholesterol levels up to 60 percent in rats. The effects lasted for weeks. This indicated that the RNA interference, or RNAi, mechanism may provide a new therapy for treating high cholesterol. Similar treatments for four nonhuman primates, conducted by a certified contract research organization, produced an average 56 percent drop in low-density lipoprotein cholesterol levels in the animals' blood, the researchers added.

The siRNA works simply by jamming the production of PCSK9—proprotein convertase subtilisin/kexin type 9—a protein that normally raises the level of LDL cholesterol—the "bad" cholesterol that tends to lead to fatty deposits inside blood vessels.

"It's very clear that eliminating this protein has cardiovascular benefits," said Dr. Jay Horton, professor of internal medicine and molecular genetics at UT Southwestern and the study's co-senior author.

The RNAi method also performed as well as cholesterol-lowering drugs currently on the market or in clinical trials, Dr. Horton said. Those medicines can provide about 20 to 50 percent drops in LDL cholesterol, but patients usually have to take maximum doses over prolonged periods.

"RNA-based drugs might provide a course of treatment for people whose cholesterol levels are resistant to current drugs, or they might be combined with current drugs," said Dr. Horton.

Working with scientists at MIT and at Alnylam, an RNAi therapeutics company in Cambridge, MA, Dr. Horton and his colleagues designed siRNAs to block the process by which DNA creates the PCSK9 protein. Normally, to make a protein, DNA's genetic code is translated to form a correspondingly coded RNA, which carries out instructions that tell the cell to make the protein.

But in the study, the lab-designed siRNA latched onto the cell's much larger RNA, preventing it from completing the PCSK9 protein-creation process.

The scientists made versions of siRNA that blocked the forms of PCSK9 found in mice, rats, nonhuman primates, and humans. Those siRNAs were then injected into normal rodents and nonhuman primates, as well as into mice that had been genetically engineered to produce human PCSK9.

The siRNAs, researchers demonstrated, increased the levels of PCSK9 to drop up to 70 percent in mice livers and 60 percent in rat livers, where the protein is originally produced. The nonhuman primates also presented with a significant drop in blood levels of PCSK9.

In addition to the drop in PCSK9, the levels of cholesterol in the blood dropped by about one-third in mice and nearly two-thirds in rats, according to researchers.

The nonhuman primates' LDL cholesterol dropped an average of 56 percent, with one animal showing a nearly 70 percent reduction, researchers said.*

Yet another new tactic in the fight against cardiovascular disease—utilizing nanoengineered molecules called *nanolipoblockers* against cholesterol buildup—is promising in early laboratory studies at Rutgers University.

In a paper published in American Chemical Society's journal *Biomacromolecules* on June 12, 2010, Rutgers researchers put forth a new way to combat clogging arteries by fighting against the way bad cholesterol causes inflammation and creates plaque buildup in blood vessels. The novel approach is the complete opposite of today's statin drug therapy, which seeks to reduce the amount of low-density lipids, or LDLs, through the human body.

---

* The research was supported by the Perot Foundation, the National Institutes of Health, the Royal Netherlands Academy of Arts and Sciences, and Alnylam Pharmaceuticals.

The Rutgers approach is designed to thwart biological processes that are generally beneficial and necessary. The principal investigator, Prabhas Mogue, an associate professor of biomedical engineering and chemical and biochemical engineering at Rutgers, remarked that vascular plaque and inflammation develop when certain forms of LDL are fought by white blood cells, which collect cellular debris and disease agents.

These scavengers, called *macrophages,* perform an essential role in keeping organisms healthy; their interaction with highly oxidized LDL molecules has quite the opposite effect, said Mogue.

Mogue added that macrophages accumulate large amounts of oxidized LDL and secrete chemicals that can damage the neighboring tissues and, ultimately, become fatty foam cells.

The doctors' approach thus is to create clusters of nanoengineered molecules that target specific receptor molecules on cell membranes and block these oxidized LDLs from attaching to macrophages.

Mogue is collaborating with Kathryn Uhrich, Rutgers professor of chemistry and chemical biology, and an expert at synthesizing biologically useful molecules at the nanoscale—anywhere from 10 to 100 nm long.

The researchers create a group of nanolipoblockers, or NLBs, which compete with oxidized LDL for a macrophage's attention. The NLBs attach to receptor sites on macrophages, cutting the accumulation of oxidized LDL by as much as 75 percent.

These particles are made of engineered organic strands or chains whose ends cluster around a central point, creating a structure known as a *micelle*. Dr. Uhrich synthesized chains of molecules with several different characteristics, including attracting or repelling water and having a positive or negative charge. As the chains assembled into micelles, Mogue tested them for how well they blocked LDL uptake, doctors reported.

This research into cholesterol and nanotechnology is not just happening in the U.S.A. Foreign firms are interested in the field as well.

An Israeli firm is producing canola oil containing microscopic micelles within which can be placed vitamins, medicines, and nutritional agents that are transported directly to the sites of absorption. Within this system, insoluble in water or fats that, under traditional conditions, would not be able to reach their sites of action, are transported there in the micelle nanotechnology.

Doctors say these micelles, which contained rapeseed oil, can enter the bile produced by macro micelle, and they contain plant sterols that compete with cholesterol for entry into the bloodstream, thereby reducing the levels.

The most interesting example of this is the use of nanotechnology in traditional Chinese medicine. Chinese physicians are working with the herb *Cordyceps sinensis,* also known as "Winter Worm" and "Summer Grass." This is an exotic medicinal mushroom known in China for centuries. Two thousand years ago, in the Himalayan Mountains of Tibet, local herdsmen noticed the unusual vigor of their yaks grazing on the herb. *Cordyceps sinensis* has been used in traditional Chinese medicine as a cure-all to heal lung, liver, heart, and kidney diseases; to treat fatigue, cancer, and male and female sexual dysfunction; to relieve pain; to enhance overall health; and to promote longevity. Due to its efficacy against a variety of health conditions, *Cordyceps* has held, and continues to hold, a highly esteemed position in the vast ranks of Chinese herbal remedies.

Most people in the West have come to know this once rare herbal medicine in only the last 20 years or so. Modern researchers are now using DNA mapping, cloning, and nanotech techniques to improve the potency of *Cordyceps*.

Marvlix is one new nanotech product generated from *Cordyceps sinensis,* developed by Elixir Industry's proprietary technology, which applies nanotechnology to a bio-enhanced extraction process.

The technology utilizes electro-potential method to extract selected molecules preferentially, on the basis of predefined molecular profiles, and then bioactivates the compounds extracted. The process completely eliminates heavy metals, contaminants, and toxic compounds commonly found in traditional herbal products, while retaining all the beneficial substances in *Cordyceps*. The firm is, on foreign web sites, touting the innovation for therapeutic use in the following areas:

- Alternative cancer treatment
- Anti-aging
- Liver health and high cholesterol
- Heart health
- Kidney health

- Lung health
- Reduction of drug toxicity

The use of nanotechnology allows Elixir's products to penetrate the cell wall to release nutrients that were lost through the traditional process, researchers said.

Active compounds are, therefore, removed one by one at the molecular level, based on predefined molecular profiles. In this way, are able to preserve the full spectrum of active elements that *Cordyceps* can provide.

Researchers are, in the main, looking for new applications for cholesterol-reducing therapy. The latest study of note is *Self-Assembled Nanoparticles of Cholesterol-Conjugated Carboxymethyl Curdlan as a Novel Carrier of Epirubicin.*[2] The purpose of this research project was to develop nanoparticles made of cholesterol-conjugated carboxymethyl curdlan (CCMC) trapping epirubicin (EPB) and determine their *in vitro* and *in vivo* potential, according to researchers. CCMC was synthesized and characterized by Fourier transform infrared spectra (FT-IR) and proton nuclear magnetic resonance spectra (^{1}H NMR). The so-called degrees of substitution (DS) of the cholesterol moiety were 2.3, 3.5, and 6.4, respectively, according to researchers.

The EPB-loaded CCMC-3.5 nanoparticles were readied by the remote loading method, the scientists said. The physicochemical characteristics, drug-loading efficiency. and drug-release kinetics of EPB-loaded CCMC-3.5 nanoparticles were easily characterized, as a result of the project.

In vitro release profiles showed that EPB release was sensitive to the pH as well as the drug loading contents, researchers said. Cellular cytotoxicity and cellular uptake were regulated by using human cervical carcinoma (HeLa) cells. EPB-loaded CCMC-3.5 nanoparticles were determined to be more cytotoxic and have a wider distribution within the cells. *In vivo* pharmacokinetics and biodistribution were investigated after IV injection in rodents. Holding out the most promise is a 4.0-fold increase in the mean residence time (MRT), a 4.31-fold increase in the half-life time, and a 6.69-fold increase in the area under the curve of EPB, which were achieved by the scientists for the EPB-loaded CCMC-3.5 self-assembled nanoparticles contrasted with the free EPB. The drug level was significantly increased in the liver at 24 and 72 hr. Nonetheless, it decreased in the heart at 8 and 24 hr when contrasted with the free EPB. Researchers believe the nanoparticles might have potential application as an anticholesterol drug delivery system due to good results *in vitro* and *in vivo.*

References

[1] Choi, Won-Bong, et al. Monitoring blood cholesterol with carbon nanotubes. *Nanowerk News,* Jan. 26, 2006, www.nanowerk.com. Also reported in the Feb. 28, 2006 print issue of *Nanotechnology.*

[2] Li, Lei, et al. Self-Assembled Nanoparticles of Cholesterol-Conjugated Carboxymethyl Curdlan as a Novel Carrier of Epirubicin. *Nanotechnology* Vol. 21, 265601; doi: 10.1088/0957–4484/21/26/265601.

For Further Reference

Illinois' AuraSense Garners $2.5M Investment for Nanotechnology-Based Therapy, 12/08/09. http://www.sciencedaily.com/releases/2006/05/060510232337.htm.

What is Cholesterol? A Definition. http://www.gebooki.com/2010/09/what-does-cholesterol-have-to-do-with-heart-disease-medicine-and-prevention-tips/.

Monitoring Blood Cholesterol with Carbon Nanotubes. *Nanowerk,* 1/26/06, http://www.nanowerk.com/news/newsid=254.php.

Synthetic "Good" Nanotechnology Cholesterol. *Nanowerk News,* Jan. 9, 2009, http://www.nanowerk.com/news/newsid=8827.php

Cholesterol and Nanoparticles, by Robert Oszakiewski, Feb. 22, 2010, http://www.nanolawreport.com/2010/02/articles/cholesterol-and-nanoparticles/.

4

Nanotechnology

Drug Development and Delivery

Drug development generally is a difficult, lengthy process. But new nanotechnology drug delivery techniques are providing hope that the creation of new elixirs to treat disease will someday become more streamlined.

"Drug development" is something of a blanket term used by medicine to define the process of bringing a new drug or device to the market. The process includes drug discovery and product development, preclinical research on microorganisms/animals and, lastly, clinical trials on humans. The process used to be called *preclinical development,* but that term is now considered archaic.

This process can sometimes be quite complex. Candidates for a new drug to treat a disease may include from 5,000 to 10,000 chemical compounds. Typically, about 250 of these will show sufficient promise for further evaluation using laboratory tests, mice, and other test animals. Typically, about ten of these will qualify for tests on humans, according to a report by Dr. H. G. Stratmann.[1]

New Chemical "Entities"

New chemical entities (NCEs) are promising compounds that emerge from the process of drug development. These compounds demonstrate promise in action against a particular biological target thought to be important in given disease. Surprisingly little is understood, at this point, about the safety, toxicity, pharmacokinetics, and metabolism of these new compounds in humans. Once a compound is discovered, the drug development team expands, brings in new personnel, (with different skill sets) to assess all of these factors prior to human clinical trials.

Another significant objective of drug development is to determine the clinical dosing for the drug and to schedule the drug for first time use in a human clinical trial. This is called the *first-in-man (FIM)* or *first human dose (FHD)* by doctors.

What is more, drug developers must establish the physicochemical properties of the new compound, including its chemical makeup, stability, and solubility. The process through which the chemical is made is streamlined so that the compound, made at the bench on a milligram scale by a synthetic chemist, can be mass manufactured on the scale of a ton or more.

Moreover, the process is examined by quality control experts for its suitability. Will the drug be made into capsules, tablets, aerosol, intramuscular injectable, subcutaneous injectable, or intravenous formulations? That is determined here. These steps are known as *chemistry, manufacturing, and control (CMC).*

Much of drug development is focused on conforming to the regulatory requirements of drug licensing authorities, which in the U.S.A. would be the Food and Drug Administration (FDA). These requirements generally consist of a number of tests needed determine the toxicities of a new compound prior to first use in a man or woman. These are legal requirements. An assessment of major organ toxicity must be performed—examining the drug's effects on the heart and lungs, brain, kidney, liver, and digestive system—as well as its effects on other parts of the body that might be affected by the drug.

Often, these tests can be made using *in vitro* methods—i.e., with isolated cells. Many tests can only be made by using lab animals, since it is only in an intact organism that the complicated interplay of metabolism and drug exposure on toxicity can be really studied.

This process of drug development does not end once human clinical trials begin. Researchers must also ensure that long-term toxicities are determined, as well as effects on systems not previously monitored, like fertility, reproduction, immune system, and so forth. The novel compound will also be tested by scientists for its capacity to cause cancer. This is called *carcinogenicity testing*.

Researchers noted that if a compound emerges from these tests with an acceptable toxicity and safety profile, and if it can further be demonstrated to have the desired effect in clinical trials, then it can be submitted to the FDA for marketing approval. This process is called a New Drug Application in the U.S.A. Most new compounds fail during drug development, generally because they have some unacceptable toxicity or they simply do not work as promised in clinical trials.

Skyrocketing Costs

Drug discovery is becoming more expensive, and it is important to look at new ways to bring forward NCEs.

Studies published in 2003 report an average pretax cost of approximately $800 million to bring a new drug to market, according to an industry report.[2] The value of improving productivity in the drug development process is faster times and better decisions. Another study published in 2006 estimates that costs vary from around $500 million to $2 billion, depending on the therapy or the developing firm.

These figures relate only to new, innovative drugs associated with an NCE. Each year, worldwide, only about 26 such drugs enter the market (Table 4.1). The development costs of generic drugs are significantly lower, *Pharmacoeconomics* notes.

Table 4.1 Number of Drugs Entering the Market

Year	No.
2009	26
0228	25
2007	18
2006	22
2005	26
2004	24
2003	26
2002	28
2001	23
2000	26
1999	33

Nanotechnology in Drug Delivery, Development

A new report by market researchers GBI Research indicates, however, that nanotechnology promises to speed up the drug development process.[3] The nanotechnology-enabled drug delivery market has "plenty of opportunities and a high level of unmet need," this report states. A number of small companies have already developed novel methods for exploiting nanotechnologies in drug delivery.

The emergence of new companies is providing licensing and merger and acquisition (M&A) opportunities for larger pharmaceutical markets. This new market is also characterized by high levels of unmet needs, according to the report. "This is primarily because development in this area is in its initial stages and has a significant number of unaddressed issues such as nanotoxicity, biological behavior and distribution pattern of nanoparticles," the report adds. "These unmet needs represent a significant opportunity for companies to explore."

The report's authors note that nanotechnology R&D funding is expected to increase significantly in the future. "Among all the major geographies, the European government is investing more in nanotechnology R&D. This suggests that there will be increased governmental support for nanotechnology R&D," the authors note. "In addition, increases in early-stage venture capital will also affect the market positively."

One place where venture capital for drug delivery is flourishing now is Israel.[4] "There's hi-tech, and then there's high-level hi-tech—and on that level, you find Israel's fledgling nanotechnology industry," reports the the *Jerusalem Post.*

The industry recently held a conference in Tel Aviv. The conference, NanoIsrael 2010, was sponsored by the Israel National Nanotechnology Initiative (INNI), an umbrella group that monitors research, development, and commercial efforts in the field by Israeli companies and university research centers.

INNI reckons that there are more than 80 large and small companies laboring in Israel's nanotech sector, along with approximately 40 academic and governmental labs, employing nearly 300 researchers and scholars.

The biggest projects to emerge so far are nanotech drug delivery devices. Prof. Reshef Tenne, of the Weizmann Institute, is leading a group that discovered and studied the inorganic fullerene-like nanospheres and nanotubes, generally termed *IF nanoparticles,* considered to be a new class of nanomaterials for drug delivery. Another firm in town is developing agents for oral drug delivery.

Israel, according to INNI, has the third-largest concentration of nanotech start-up companies in the entire world, surpassed only by California's Silicon Valley and the Boston technology corridor.

An INNI research report shows that the Technion University employs 119 nano-researchers, followed by 55 researchers at Tel Aviv University, 47 at Ben-Gurion University of the Negev, 43 at the Weizmann Institute of Science, 39 at the Hebrew University of Jerusalem, and 30 at Bar-Ilan University.

Overall, the number of nano-researchers in Israel has doubled since 2002. According to the *Jerusalem Post,* most of the new medical applications being developed in Israel use nanotechnology.

Prof. Arie Zaban, of Bar-Ilan University, is very optimistic about the possibilities for Israel and nanotech drug development. "We have a lot of competition, especially from the 'tigers' of the Far East, such as South Korea and Singapore," he told the *Jerusalem Post.* "But we're a tiger of the Middle East. Our high-tech future is riding on nanosuccess." Established companies are also ramping up in new drug delivery methodologies using nanotechnology.

Turnaround biotech success story Elan Corp. (NYSE: ELN) has made some savvy strategic moves since skirting bankruptcy after a number of financial scandals a few years ago. Analysts say its best decision was to not sell off its drug delivery unit, something that almost occurred in its search for much-needed money. That drug-delivery division recently reported that its proprietary nanocrystals technology gained its first commercial success with Johnson & Johnson's nanocrystal-based version of its immunosuppressant drug, Rapamune (sirolimus). This usage was cleared by the U.S. Food and Drug Administration (FDA) in 2009. Elan recently also announced a new partnership with Swiss giant Roche, under which it will provide Roche with formulation services and technology in exchange for research revenues, development milestones, and royalties on sales of any product incorporating nanocrystal technology.

Analysts also note that Elan is highly valued right now, but the technology cannot boost its valuation to the same degree that Abraxane lifted APPX. Another assured commercial success in nanotech drug delivery development is Flamel Technologies (NASDAQ: FLML).

The firm's proprietary Medusa technology is leading the way. Back in 2004, the company suffered some setbacks when Bristol-Myers Squibb suddenly pulled out of a development deal on its lead product, Basuli, which uses a nano-encapsulation approach that has been validated by FDA. Flamel's technology utilizes the unique size and properties of the nanoscale to be more accurate in drug delivery and dosage.

Another firm with a nano-formulated product already on the market is Novavax (NASDAQ: NVAX). The product is Estrasorb, a topical emulsion for use in estrogen therapy and other applications. The firm received FDA approval last year, based upon the product's proprietary micellar nanoparticulate technology. The company is actively pursuing multiple drug candidates formulated with the same micellar nanoparticle delivery technology.

Futuristic Drug Delivery Mechanisms

But perhaps the most promising of all the nanotechnology drug delivery mechanisms is the most high-tech of them all, according to reports in *Nature Magazine* and other scientific media.

Intravascularly injectable nanovectors are the major class of nanotechnological devices of interest for use in cancer, according to industry reports. The planned use is for the *in vivo,* noninvasive visualization of molecular markers of early stages of disease; the targeted delivery of therapeutic agents, with a simultaneous, substantial reduction of side effects; and—by a combination of the first two—the interception and containment of lesions before they reach the lethal or even the malignant phenotype, with minimal or no substantial loss of quality of life.

Liposomes are an archetypal, simple form of a nanovector. These technologies use the over expression of fenestrations in cancer neovasculature to increase drug concentration at tumor sites, industry reports note. Liposome-encapsulated formulations of doxorubicin were approved ten years ago for the treatment of Kaposi's sarcoma and are now used against breast cancer and refractory ovarian cancer, according to industry reports. Researchers are refining liposomes and applying them to more cancer indications.

They are at the leading edge first of an ever-growing number of nanovectors under development for new, more efficacious drug-delivery modalities, according to *Nature.*

Several new nanoparticles for the enhancement of MRI contrast have been used clinically and in research protocols, according to a report in *Nature.* These nanotech delivery devices include gadolinium-based, iron oxide-based nanoparticles and multiple-mode imaging contrast nano-agents that combine magnetic resonance with biological targeting and optical detection, *Nature* reports.

Low-density lipid nanoparticles, what is more, have already been used to bolster ultrasound imaging. For every clinical modality it is possible to create nanoparticles that furnish signal enhancement, when merged with bimolecular targeting capabilities.

Nanovectors, generally, have a tripartite constitution, presenting a core constituent material, a therapeutic and/or imaging payload, and biological surface modifiers, which enhance the biodistribution and tumor targeting of the nanoparticle dispersion.

A significant clinical advantage sought by the use of nanovectors over simple immunotargeted drugs is the specific delivery of large amounts of therapeutic or imaging agents per targeting biorecognition event, according to *Nature.*

Targeting methods that have been examined range from covalently linked antibodies to mechanisms. Researchers report that multifunctionality is the fundamental advantage of nanovectors for the cancer-specific delivery of therapeutic and imaging nanotech agents. Main functionalities include the avoidance of biobarriers and biomarker-based targeting, and the reporting of therapeutic efficacy, according to *Nature.*

Thousands of nanovectors are currently under study by medicine, according to *Nature.* By combining them, in a systematic way, with preferred therapeutic and biological targeting moieties, it might be possible to obtain a large number of novel, personalized therapeutic agents, researchers believe.

Novel mathematical models are needed to secure the full import of nanotechnology into oncology, however. Nanoparticles are also found as delivery devices in personal care products such as cosmetics and sunscreens.

According to the Cosmetic, Toiletry, and Fragrance Association, nanoscale grades of titanium dioxide and zinc oxide have been approved by FDA for use as sunscreens since 1999 and are also used in cosmetics. Nanoscale formulations of these molecules are clear instead of opaque, enhancing their usefulness in products applied to the skin.

Cosmetics are sold without the premarketing scrutiny that the FDA gives drugs, although the agency can take regulatory action if problems are discovered after marketing starts.

References

[1] Stratmann, H. G. Bad Medicine: When Medical Research Goes Wrong. *Analog Science Fiction and Fact,* Sep. 2010, (9): 20, www.analogsf.com.

[2] *Pharmacoeconomics.* 20 suppl. 3: 1–10 PMID 12457421.

[3] The Nano-Enabled Market Represents Significant Opportunities and High Level of Unmet Needs. GBI Research, Nov. 8, 2010, www.gbiresearch.com.

[4] Shmah, David. Nanotechnology Applications Could Be Israel's Next Move. *The Jerusalem Post,* Nov. 8, 2010.

Recommended Reading

DiMasi, J. The Value of Improving the Productivity of the Drug Development Process: Faster Times and Better Decisions. *Pharmacoeconomics* 20 suppl. 3: 1–10 (2002). PMID 12457421

DiMasi, J., R. Hansen, and H. Grabowski. The Price of Innovation: New Estimates of Drug Development Costs." *J Health Econ* 22 (2) (2003): 151–85. doi: 10.1016/S0167–6296(02)00126–1. PMID 12606142

Adams, C., and V. Brantner. Estimating the Cost of New Drug Development: Is It Really 802 Million Dollars? *Health Aff* (Millwood) 25 (2) (2006): 420–8. doi:10.1377/hlthaff.25.2.420. PMID 16522582.

GBI Research. The Nano-Enabled Market Represents Significant Opportunities and High Level of Unmet Needs. Nov. 8, 2010, http://www.gbiresearch.com/Report.aspx?ID=Nanotechnology.

5

Molecular Imaging and Diagnostics

A new era in imaging—and diagnostics—is emerging, courtesy of nanotechnology.

According to researchers at Stanford University, innovations in the materials used to manufacture semiconductor chips are opening new vistas for medical imaging, enabling physicians to locate diseased tissue with more precision than previously believed possible.

The technology is called *quantum dots (Qdots),* the Stanford researchers indicate. The technology is based on small, nanocrystal creations made from the same materials as computer chips. These structures are injected by radiologists into animals, and they help produce high-resolution, multicolored images of single molecules. The structures flow throughout the animals' body, pinpointing diseases or likely diseased tissues.

"Qdots are the perfect merger of nanotechnology with molecular imaging," said Sanjiv Sam Gambhir, who heads the Stanford University School of Medicine's molecular imaging program, in an interview for the university's web site. "Nanotechnology will be used initially in animals and soon humans to image molecular and cellular events."[1]

Gambhir reckons these nanocrystals structures are like "molecular detectives."[1] These nanoparticles act as if they are droplets of free electrons which exhibit fluorescence when stimulated with the proper light energy.

The Qdots are just a few nanometers in diameter—about a billionth of a meter. The dots are bigger than many molecules but much smaller than the cells themselves. These nanostructures are manufactured in nanofabrication facilities and have been made to function with long-term visualization properties. Several companies are beginning to sell Qdots.

Ultimately, researchers believe more effective methods will emerge to diagnose and treat cancer and other fatal diseases. Some Qdots have even been created with bits of protein attached to them that allowed them to zero in and connect to the unique proteins on certain types of cancer cells.[2] These dots even emit a specified color of light when stimulated (Fig. 5.1).

"What this technology allows that can't be done now is multiplexing—having signals of different colors letting us image many things simultaneously," said Gambhir.[1]

The problem with available imaging technologies is that physicians can only inject a single probe at a time, which emits one signal at a time.

"If I want to know 30 things about a cancer at once, I would have to inject one probe then wait for the answer from that one. Then inject another probe and wait. And so on. The multiple signals of Qdot technology are like having a colorful barcode,"

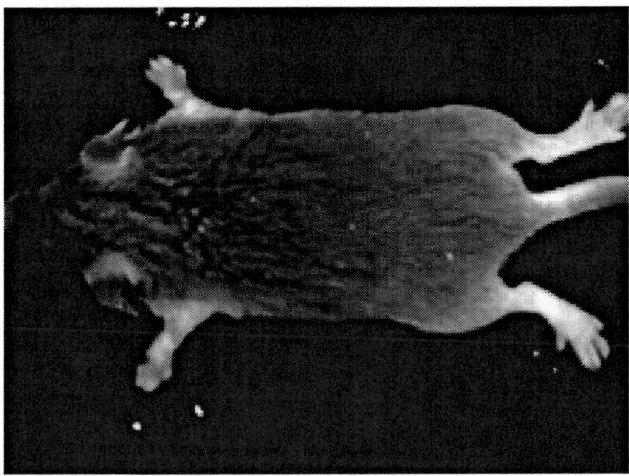

FIGURE 5.1 Researchers injected quantum dots subcutaneously into a mouse's left paws. Image Courtesy Stanford Medical School.

explained Gambhir, who has a $25 million grant from the National Cancer Institute to develop these technologies.[1]

According to the NCI, Gambhir isn't the only radiological genius working in this emerging field.[3] That means that even more promising research results are expected soon from all over the U.S.A.

The recent history of modern medicine demonstrates that important advances in treating diseases have generally followed the development of innovative ways to better see within the body. This is especially true in oncology.[3]

The debut of computed tomography (CT) imaging, for instance, provided medicine with images of developing tumors in sharper detail than was available with X-rays. This gave cancer doctors a way of localizing tumors before surgically resecting them as well as a preview of whether a particular therapy was prompting a tumor to shrink.

Similarly, magnetic resonance imaging (MRI) provided still greater anatomical detail, while the development of positron emission tomography (PET) gave both cancer researchers and oncologists the ability to monitor a tumor's metabolic activity and, as a result, an even quicker way of assessing the effectiveness of therapy.[3]

These imaging technologies, however, share from a similar shortcoming. They are not sensitive enough to pinpoint the smallest tumors that can be effectively treated by doctors. What is more, most imaging methods generate static imagery, i.e., snapshots of a tumor at one particular moment in time. These images do not reveal all that much about dynamic events, like the binding of a new drug to a particular tissue.

Nanotechnology, however, might find a way to make that leap in capacity that will not only change medicine's approach to cancer therapy but may also lead to completely new pathways for discerning and treating cancer, according to researchers.

"The promise of nanotechnology for cancer imaging is such that we have little doubt that it will lead to far more sensitive and accurate detection of early stage cancer," said Adrian Lee, PhD, an associate professor of medicine who specializes in translational breast cancer research at Baylor College of Medicine, in an interview with *Nanotech* 2006 Vol. 2.[4] "I also believe that we are just at the beginning of the process of applying nanotechnology to the problems of imaging cancer. I have confidence that as the oncology and physical sciences communities continue to find common scientific ground, there are going to be some surprising advances that will come of this work. These efforts will blur the boundaries of what we call detection."

Lee and other researchers at Baylor, like chemistry professor Lon Wilson, PhD, have been working on a project funded by the National Cancer Institute to determine how to effectively use unique nanoscale contrast agents for the MRI made from gadolinium and iron, two kinds of atoms that seemingly resonate with the influence of magnetic energy. Researchers have encased these materials within carbon nanotubes. "We have good evidence that these new contrast agents have the potential to give us a big boost in imaging sensitivity, but how exactly we'll use these nanotube-based agents and what role they will play in therapy is still an open question that we're going to work to answer," Lee told Nanotech.[4]

Another researcher, Jeff Bulte, PhD, associate professor of radiology at Johns Hopkins University, told *Nanotech* that there is no question as to how nanoparticle-enabled imaging can help bolster cancer therapy.[4]

Collaborating with Carl Figdor, PhD, and his coworkers at the Radboud University Medical Center in Amsterdam, Dr. Bulte has been experimenting with the use of iron oxide nanoparticles to monitor exactly how dendritic cells move through the body. These dendritic cells are thought by medicine to trigger immune responses that may kill tumors.

Doctors injected these cells into a patient's lymph nodes. After labeling dendritic cells with magnetic nanoparticles and tracking them using MRI, the researchers told *Nanotech*, they discovered that they were successful only 50 percent of the time. But the nanoparticles are increasing the efficacy of the therapy. "We can use a widely available imaging method, MRI, to ensure that we've accurately delivered therapeutic cells to the exact spot where they can do their job," Bulte told *Nanotech.*

But there's more. Doctors have developed multifunctional nanodevices designed to be both imaging agent and an anticancer therapy.[4]

According to James Baker, Jr., MD, director of the Michigan Nanotechnology Institute for Medicine and Biological Sciences and director of an NCI-funded Cancer Nanotechnology Platform Partnership team, he and his colleagues have been pushing forward with a research project created to develop tumor-targeting dendrimers that have an imaging agent and therapeutic agent.[3] Baker's collaborators developed a dendrimer connected to a fluorescent imaging agent and paclitaxel, the anticancer drug, and demonstrated that this innovative agent can find tumor cells and kill them, instantly.[3]

Other so-called "platform partnership" teams, headed by Kattesh Katti, PhD, of the University of Missouri-Columbia; Panos Fatouros, PhD, of the Virginia Commonwealth University; Miqin Zhang, PhD, of the University of Washington; Allan Oseroff, MD, PhD, of the Roswell Park Cancer Institute; and Paras Prasad, PhD, of the State University of New York in Buffalo, are simultaneously creating multifunctional nanoparticles for imaging and therapeutic applications.[3]

Scientists note that imaging techniques work on the same basic principle, i.e., a form of energy is beamed into the body, where that energy then interacts with the body's molecules. A detector, created just for that particular type of imaging machine, then records the interactions, which enables a radiologist to read the resulting image and make a diagnosis.

A CT scanner, for example, works by sending focused X-rays into a body, which are then deflected off or pass through the tissues and bones. A special detector measures the intensity of the X-rays passing through the body. A computer the reformats that data to create an image. For optical imaging, the form of energy is an infrared-style light. For magnetic resonance imaging, a magnetic field is combined with radio waves to stimulate all of the water molecules in the body. The way in which those water molecules relax provides doctors detailed structural information when analyzed by a computer.[3]

Sometimes, it imaging is even more complicated. Some imaging methods make use of energy that is injected into a body and record how that energy passes out of the body. PET imaging, for example, works with energy broadcast by radioactively labeled molecules that get grabbed up by cells. This is where the above-mentioned quantum dot technology developed by Gambhir comes into play.[4]

There is, however, one engineering concept, known as *signal to noise,* that shows how all of the above scientists face a huge challenge in their work. A signal-to-noise ratio refers to the idea that the data collected by an imaging technology is a mixture of the diagnostically helpful signals that proclaim there is a tumor in one location and background noise elsewhere.[3] The higher the ratio of signal to noise, the easier it is to make sense of an image.

To discern the signals from the noises, radiologists have developed imaging contrast agents, some of which were described above. These molecules have physical characteristics that increase the strength of the signal coming out of the body. MRI contrast agents containing the element gadolinium, to illustrate the problem, alter the magnetic field in the body, which boosts the strength of the MRI signal.

Meanwhile, optical imaging contrast agents behave as light-amplifying devices, which send out light signals strong enough, and of the correct frequency, to pass through tissues and skin. The trick to successful utilization of these agents is to concentrate them in tumors.

In this instance, nanoparticles can play an enabling role. "Nanoparticles are ideal for creating imaging contrast agents because of two properties," said Lily Yang, MD, PhD, an assistant professor of surgical oncology research and a member of the Emory-Georgia Institute of Technology Nanotechnology Center for Personalized and Predictive Oncology, in the interview with *Nanotech.* "We can design them so that they are very bright when imaged, and second because we can attach various targeting molecules to their surfaces and achieve a high concentration of the imaging agent at the tumor."

Nanotech in Imaging: A Norm of the Future

Medicine is trying to integrate these cutting-edge technologies into the everyday training of new doctors. This will ensure that, in the long term, nanotechnology becomes the norm in imaging.

At Northwestern University's medical school, a National Cancer Institute (NCI)-funded program called "Cancer Nanotechnology in Imaging and Radiotherapy" is actively recruiting residency candidates who have earned their MD and want to specialize in this emerging form of radiology.[4]

The residency program is a full-time training regime, lasting two or three years, and includes laboratory and didactic research components. As an incentive, the program pays a higher-than-average salary for residents—$68,000 per year, with another $20,000 for travel expenses. To be eligible, students need to have completed a standard radiology residency already. The program is being offered at the Robert H. Lurie Comprehensive Cancer Center at Northwestern, but there is strong participation from the University of Chicago's department of radiology and cellular oncology as well.

What's Next for Nanotech Diagnostics?

Plenty.

According to the *Technical Proceedings of the 2006 Nanotechnology Conference and Trade Show,* a handful of compelling technologies in laboratory development are poised to make breakthroughs in clinical settings in the coming years.[5]

Research abstracts published at that conference point to the future of nanotechnology diagnostics.[4] To wit,

- *A highly sensitive fluorescent immunoassay based on avidin labeled nanocrystals.* The authors were K.-K. Sin, C. P.-Y. Chan, T.-H. Pang, M. Seydack, and R. Renneberg of the

Hong Kong University of Science and Technology. HK Researchers developed a g/L mouse, and immunoglobulin G was created, which was lowered by a factor of 2.5 to 21 contrasted with the immunoassay using commercial labeling systems, i.e. FITC and peroxidase. What is more, the sensitivity was 3.5- to 30-fold higher, researchers said.

This study demonstrates that the fluorescent nanocrystals combined with the avidin-biotin technique can enhance the assay sensitivity and achieve a lower limit of detection without requiring long incubation times as in enzyme-based labels. The fluorogenic precursor fluorescein diacetate (FDA) was applied as labels to enhance assay sensitivity in a previous study.

Each FDA nanocrystal can be changed into $\sim 2.6 \times 10^6$ fluorescein molecules, which is useful for improving sensitivity and limits of detection of immunoassays. NeutrAvidin, doctors said, was adsorbed on the surface of the FDA nanocrystals that were coated with distearoylglycerophosphoethanolamine (DSPE) modified with amino(poly(ethylene glycol)) (PEG(2000)-Amine) as a proper interface for coupling biomolecules.

- *New Applications of Nanoparticles in Cardiovascular Imaging,* by R. R. Sharma, FAMU-FSU College of Engineering, U.S.A. Nanotechnology works at the atomic, molecular, and macromolecular levels, including imaging. Recently, four areas have emerged in cardiovascular imaging.

  Targeted therapeutics include (1) delivering drugs where they are needed; (2) tissue engineering to replace defective valves, damaged heart muscle, and clogged blood vessels; (3) molecular imaging using "smart" imaging agents in targeted therapeutics and molecular imaging; and (4) biosensors and myocardial diagnostics, including a number of new approaches of nanoparticles, i.e., dendrimers, liposomes, polymer delivery molecules, cantilevers, nanoscaffolds, nanofibers are potential candidates in cardiac visualization. The extracellular matrix plays significant role by chemokines, cytokines, and growth factors. The main focus of this paper is on the limitations of these emerging techniques and a new possibility of MRI visualization of mice cardiac atheroma by superparamagnetic iron-oxide gadolinium-apoferritin (SPIOA), myoglobin (SPIOM), for targeted functional and molecular imaging of atherosclerosis. These emerging techniques furnish a way to track functional and structural changes in the myocardium as well as heart tissue.

- *Preparation of Aggregation Stable Block Copolymer Nanoparticles for Simultaneous Drug Delivery and Imaging,* by M. E. Gindy, and R. K. Prud'homme of Princeton University. According to the authors, a preformed imaging agent, dodecanethiol capped gold, which is 1 to 3 nm, and a fluorescent molecule, 7-amino-4-methyl coumarin, can be dissolved in a water-miscible solvent such as tetrahydrofuran and subsequently impingement mixed against an opposing water stream using the confined impingement jets (CIJ) device.

The two incoming jets provide a region of high intensity micromixing where solute supersaturation is rapidly gained, and precipitation of the drug, imaging agent, and block copolymer happens without diffusional limitations.

The result: the formation of polymer protected nanoparticles containing both the drug compound as well as the imaging agent. This process yields nanoparticles with controlled size, narrow particle size distributions, high encapsulation efficiencies, and long-term stability.

What is more, the amphiphilic polymer-based delivery mechanisms may be used for parenteral, nasal, subcutaneous, intramuscular, aerosol, and oral administration.

Doctors have also developed a nanoparticle formation process based on rapid micromixing to produce high supersaturations, which they call Flash NanoPrecipitation.

This introduces a new approach which rapidly expands the application of the Flash NanoPrecipitation process to pre-formed solutes in connection with organically dissolved molecules, and enables the formation of nanoparticle formulations with limited recrystallization, and stable size distributions.

Commercialized Diagnostic

Though most of the above-cited work is lab-based and relatively early stage, at least one company has developed a marketable nanotechnology and is selling it to oncologists. That's the TriLite Technology developed by Crystalplex Corporation. The technology is used to create nanocrystals and nanoclusters for imaging and diagnostics. TriLite alloyed nanocrystals are of relatively uniform size: approximately 6 nm, according to the developer.[5]

These alloyed nanocrystals can be produced in stable blue and blue green colors, i.e., less than 525 nm. These crystals can also be made in the far red region of the light spectrum, e.g., greater than 660 nm, with little loss in stability.

The developer indicates that TriLite nanoclusters are aggregates of 8 to 12 individual nanocrystals. The nanoclusters are 40 to 50 nm in size and are functionalized on the surface with carboxyl groups using a proprietary Crystalplex technology. Diagnostic products are available in many sizes and colors and can also have nearly any biological probe bound to them.[5]

Clearly, a new era in imaging—and diagnostics—is coming of age, courtesy of nanotechnology.

References

[1] Stanford Researchers Develop Tool that "Sees" Internal Body Details 1,000 Smaller. Stanford University Medical Center, http://med.stanford.edu/news_releases/2008/march/raman.html.

[2] NCI Funds New Nanotechnology Cancer Center at Stanford, http://www.ed.stanford.edu/news_releases/2006/march/nanotech.html.

[3] Cancer Nanotechnology in Imaging and Radiotherapy Program, Northwestern University Medical School, http://janus.northwestern.edu/R25/index.php.

[4] Alper, Joe. Nanotech Driving Big Advances in Cancer Imaging. *Nanotech* 2006, Vol. 2, April 23, 2006, http://www.nsti.org/procs/Nanotech2006v2.

[5] TriLite Technology, Crystalplex Corporation (USA), http://www.crystalplex.com.

6

Creating Customized Bones

Nanobones

You're in a dreadful car crash, and your lower right leg, the tibia bone, is damaged, seemingly beyond repair. In the past, this would have been catastrophic—a life-changing event. No amount of wiring and metal grafts would do, or bear the weight of your body, like a real bone. But in the coming years, nanotechnology will be used to replace that missing part of your lower limb, as medical scientists are cultivating replacement nanobones in labs. And accident victims won't be the only ones benefitting from this dramatic technological transformation. Folks who need new teeth or hip replacement surgery will have this kind of therapy available as well.[1]

Current technologies, such as bone grafts, are helpful for some. Doctors lift parts of bones from the patient's body, as home-grown bone is something the body is less likely to reject. However, getting a graft entails another incision, more pain, and increased risk of surgical complications.

Nanotech developments in bone replacement are emerging all over the world, promising to remodel the entire profession of orthopedic surgery.

Laura Zanello, an assistant biochemistry professor at the University of California, Riverside, is using lab-grown nano-bone, grown from the bone cells of rats, to substitute for missing bone fragments. Zanello's team has developed a technology in which bone cells grow onto tiny scaffolds built of carbon nanotubes. These structures are no more than a few nanometers in diameter. Once this technique is perfected, the nanotubes will be layered with the patient's bone cells and will fit precisely into a gap in a damaged bone. The bone cells merge with the surrounding bone, just like a conventional graft.[1] Carbon is bio-friendly, and the bone would be grown from the patient's own cells, making it unlikely that the body would reject the nanotech bone graft.

The key innovation was producing nanotubes using less-toxic metals, said Zanello. Metals are toxic to most living cells. Bin Zhao, a researcher at Oak Ridge National Laboratory, produced nanotubes made of metals that are purer than previous versions, giving the team less-toxic scaffolds upon which to place the bone cells.

"The most fascinating part was that they will not only grow and proliferate, but they secrete a bone matrix," said Zanello.[1] The matrix allows the cells to fuse with extant bone, she added.

Across the country, at the University of Arkansas, researchers are working on a related problem. A team of researchers has found a new and relatively cheap way to make a nanowire coating for titanium surfaces used in bone implants. The nanowire scaffolds are being employed to make more effective surfaces for hip replacement, dental reconstruction, and vascular stenting. Medical scientists can control the length, the height, the pore openings, and the pore volumes within the nanowire scaffolds.[2]

Researcher Z. Ryan Tian, assistant professor of chemistry and biochemistry at the University of Arkansas, is leading this effort. Tian took on this particular medical problem because he noted that bone implants don't last forever, and an effective replacement technology is needed for the long term.

"Reconstructive bone surgeries, such as hip replacements, use titanium implants. However, muscle tissue may not adhere well to titanium's smooth surface, causing the implant to fail after a decade or so and requiring the patient to undergo a second surgery," said Tian.[2]

Tian and his team produced a nanowire-coated joint and placed it in mice. After a month, the researchers found that tissue had actually adhered to the joint. "We saw beautiful tissue growth—lots of muscle fibers," Tian said. "We've added one more function to the currently-in-use titanium implant."

There are a number of possible uses for this technology. Medical researchers using this nanotech can control the size and shape of the pores in the nanowire scaffold. What's more, the material could also be used on stents in patients with coronary artery disease and in potential stroke victims. Regular stents often become reclogged with fat after implantation by a surgeon. The drug-eluting stent, another technology used to treat that particular problem, is made of a polymer mixed with the drugs. But the coating is biodegradable and does not maintain function for a long time. A nanowire coating could be used to carry drugs that would help keep the arteries clear over a long period of time, said Tian.[3]

Across the world, in Australia, researchers note that 50,000 Australians each year end up with metal knee and hip replacements.[4] Researchers at Murdoch University, Perth, have created a nano material that mimics the structure and composition of real bone, and which may make artificial titanium implants archaic.

Old fashioned artificial implants, made of titanium metal, are surgically inserted. These implants are often the sites of infections in patients—dangerous post-surgical complications. Murdoch scientist, Dr. Gérrard Poinern, has created a new formulation of the mineral powder made of the main component of bone—a ceramic called hydroxyapatite (HAP).

"Hydroxyapatite is chemically similar to the mineral component of bones and hard tissues in mammals, which gives bones their strength," said Poinern.[4] "The 'nanobone' material could potentially replace the traditional titanium joint with a nanobone plate that is more widely accepted by the body. Nanobone implants have better capacity to interact with living tissue, allowing the body to repair itself much faster as it recognizes similar nano material and tries to grow into it."

Poinern added, "We know bone is made up of little rods of hydroxyapatite, so we have manufactured a new version of HAP that can be shaped into any form, including current implant designs. What's really exciting is that, because it's a powder, it can be shaped into screws, plates, and any other shape required. And on top of that, it also opens the way for us to experiment with combining this nanobone with other substances at the nano scale to maybe increase strength or produce other novel attributes."[4]

Poinern said that nanobone technology may also be dosed with antibiotics, which are released when bacteria enter the site of the joint replacement, thus decreasing chances of the implant being rejected by the body.

The president of the Australian Orthopedic Association, John Batten, said that the science behind these innovations may one day become a great tool for orthopedic surgeons. "It has great potential for prosthesis, for implants, and for fracture healing and other things we do to the skeleton," said Dr. Batten.

Poinern's team relied on millions of spheres measuring 37 nm long. (A nanometer is one millionth of a millimeter.) A human hair is 80,000 nm wide, by comparison.

Researchers on Poinern's team, the Murdoch Applied Nanotechnology Research Group (MANRG), combined chemicals like calcium nitrate, ammonium hydroxide, and potassium hydrogen phosphate for the project. When heated with ultrasound waves, they formed tiny mineral spheres.[4]

There's still work to be done on this technology, however. "Part of the journey ahead is to figure out its applications and to work with partners in the medical device industry and surgeons to assess what exciting new products could be manufactured," Poinern said.

Back in the United States, one company that works on nanobone technologies is generating a lot of enthusiasm from investors.[5] Malvern, PA-based Orthovita, Inc. (NASDAQ: VITA), an orthobiologics and biosurgery company, is constantly making headlines with its nanobone innovations. On Oct. 6, 2010, the company announced that Dr. Hyun W. Bae, MD, of the Spine Institute at Saint John's Health Center in Santa Monica, CA, published an abstract entitled, "Correlation of Early Pain and Long-Term Functional Results from a Multi-Center, Prospective, Randomized, Controlled FDA-IDE Vertebroplasty Trial." It was presented at the 25th Annual Meeting of the North American Spine Society (NASS). The convention was held October 5–9, 2010, at the Orange County Convention Center in Orlando, FL. The abstract was selected for presentation during the friday "Best Papers" session.

The presentation during the fall of 2010 at NASS came after a decade of preclinical and human clinical research on Cortoss Bone Augmentation Material. Since 1999, the research has been highlighted in 49 podium and 33 poster presentations at a variety of national and international specialty society meetings, and published in 18 peer-reviewed medical journals. "The sheer number and nature of scientific presentations over the years reflect our commitment to develop and clinically evaluate this unique material, for which we plan to seek regulatory clearance in additional indications," said Maarten Persenaire, MD, chief medical officer of Orthovia.[5]

Orthovita is an orthobiologics and biosurgery company that develops and markets novel medical devices. The company's orthobiologics platform produces products for the fusion, regeneration, and fixation of human bones. The firm's biosurgery platform provides products for controlling intraoperative bleeding, or hemostasis. The firm's fusion and regeneration products are based on the proprietary VITOSS™ Bone Graft Substitute technology. Another technology developed by the firm, CORTOSS™ Bone Augmentation Material, is an injectable polymer composite that actually mimics the structural characteristics of human bone. The company also makes VITAGEL™ Surgical Hemostat, a collagen-based matrix that controls bleeding and facilitates healing, and VITASURE™ Absorbable Hemostat, a plant-based product that can be deployed rapidly during surgery.[5]

Dr. Bae's abstract was based on clinical data from Orthovita's multi-center, prospective, randomized, investigational device exemption (IDE) study. The safety and efficacy of Orthovita's Cortoss™ Bone Augmentation Material was contrasted with a polymethylmethacrylate (PMMA) control in the treatment of vertebral compression fractures. Dr. Bae was a clinical investigator for the study.

Dr. Bae reported that the safety and effectiveness of Cortoss was found to be "non-inferior" to the control technology. What is more, key findings from the 256-patient clinical trial include the statistically significant benefits for the Cortoss patient group.[5]

To wit,

- Early patient outcomes assessed at three months, with 86.6 percent of Cortoss patients and 75.0 percent of PMMA patients achieving "successful pain relief" as measured by the Visual Analogue Pain Scale (VAS). Cortoss patient group experienced a "statistically significant" benefit in pain relief over the PMMA group.

- Long-term patient outcomes were assessed at 24 months, with 96.7 percent of Cortoss patients and 88.4 percent of PMMA patients maintaining or improving function as measured by the Oswestry Disability Index (ODI). The Cortoss patient group experienced a statistically significant benefit in function success over the PMMA group.

The Cortoss population experienced measurable benefits over the PMMA cohort in the following outcomes, Dr. Bae reported:

- A 43.4 percent reduction in adjacent-level fractures in Cortoss patients treated for a primary fracture at one level, as measured at the end of the 24-month follow-up period.
- An average of 30 percent less material was required to achieve proper fracture fill in Cortoss patients.

"With all of the discussions surrounding vertebral augmentation that are taking place today, this level I study provides important evidence that should be included in the debate," said Dr. Bae. "The results show that the majority of properly selected vertebroplasty patients experience early, substantial, and sustained reduction in pain and improvement in function. Furthermore, the results show that the improved material characteristics of Cortoss as compared to PMMA do lead to measurable improvements in outcomes."

There were other developments at the company during the fall of 2010 at Orthovia as well. The company received 510(k) clearance from the U.S. Food and Drug Administration to market VITOMATRIX Bone Graft Substitute as a bone grafting material in certain dental procedures.

VITOMATRIX is a resorbable, highly porous, synthetic scaffold that utilizes the core proprietary technology from the VITOSS™ Bone Graft Substitute product line. VITOMATRIX is indicated for use in procedures to fill, augment, or reconstruct periodontal or bony defects of the oral and maxillofacial region. Approximately 375,000 procedures are performed domestically each year where VITOMATRIX can be used, the company said.[6]

Orthovita is evaluating potential commercial partners to distribute the product or license its underlying technology, the company said in a news release.[6]

During mid August 2010, the company also published a study demonstrating the efficacy of its vertebroplasty technology.[7,8] The study was the first, large multicenter prospective study comparing the efficacy of nanotechnology-based vertebroplasty to traditional care in the treatment of vertebral compression fractures.

The 202-patient randomized study demonstrated that vertebroplasty resulted in greater pain relief than conservative treatment at all time points.[8] Researchers reported no serious complications or adverse events.

"Vertebroplasty versus Conservative Treatment in Acute Osteoporotic Vertebral Compression Fractures (Vertos II): An

Open Label Randomized Trial," authored by Drs. Klazen and Lohle et al., described the methods and findings of a 202-patient study conducted in the Netherlands and Belgium between October 1, 2005, and June 30, 2008.

Vertebroplasty is a minimally invasive procedure involving the under-the-skin injection of nanotech-based bone cement into a fractured vertebral body, effectively reducing pain. Orthovita received FDA clearance in June 2009 for Cortoss™ Bone Augmentation Material, a synthetic biomaterial, for use in vertebroplasty as an alternative to polymethylmethacrylate (PMMA) bone cement. Conservative care included optimized analgesic regimens, bed rest, and physical therapy. According to the authors, "Decrease in VAS score after vertebroplasty was significantly higher than with conservative treatment at all time points." What's more, the vertebroplasty patients, who had significantly more disability and lower quality of life scores at baseline, showed significantly faster and greater improvements than conservatively treated patients for these parameters. The authors concluded, "Pain relief after the procedure is immediate, sustained for one year and is significantly better than that achieved with conservative treatment and at acceptable costs."[8]

Back overseas, Chinese researchers are also working with nanotechnology on bone-related disorders. Doctors in China have recently treated 18 bone disease patients by implanting a new type of artificial bone.[9]

The technology, also called "nanobone," was developed by a group of scientists from Qinghua University, Beijing, China, headed by Cui Fuzhai. The team claims that this novel type differs from other iterations because this type of material disintegrates over time, unlike other nanotechnology innovations.

Back in the States, at least one firm is riding the wave of nanotechnology not through technological innovation but through marketing savvy. This shows the maturation of this still young industry.[10] Soteria, Inc., of Dedham, MA, has licensed nano biomaterial technology from Competitive Technologies, Fairfield, CN. Soteria is going to manufacture, use, and sell products using CTT's nanotechnology bone biomaterial for applications related to the human spine.

The Dedham, MA-based firm is creating a marketing program to develop and obtain government agency regulatory approval for the marketing and sale of the biomaterial. The technology was developed by CTT and the University of South Carolina Research Foundation (USCRF). It consists of an injectible calcium phosphate-based biomaterial and comes from the pioneering work of Dr. Brian Genge, a research professor in the Department of Chemistry and Biochemistry at the University of South Carolina.

The licensing agreement limits the use of the technology for human spinal therapy. The technology, however, can be applied to a variety of additional human orthopedic procedures and is available for licensing for craniofacial reconstruction, hip replacement, as well as dental and veterinary uses. The bone biomaterial technology is based on Dr. Genge's development of nanoparticle calcium phosphate materials. The bone cement is a flowable, moldable paste that conforms to and *interdigitates,* or mingles, with the host bones and rapidly hardens itself *in situ*, forming a solid bone-like structure capable of stabilizing fractured bone within ten minutes. The cement remodels into bone over a period of time. The technology is suitable for weight-bearing and non-weight-bearing bones, and is both machinable and drillable by surgeons.

"The agreement with Soteira to develop products using USCRF's innovative nanoparticle biomaterial technology in the $2.5 billion spinal products market is an excellent example of CTT fulfilling unsatisfied market needs from our portfolio of technology solutions," said D. J. Freed, PhD, CTT's president and chief executive officer.

According to Larry Jasinski, Soteira's president and CEO, his company is creating a development and complete testing plan, hoping gain swift FDA approval and achieve "rapid commercialization of this technology for use in treating degenerative diseases of the spine."[10]

But, even though the industry is maturing somewhat, that doesn't mean the innovations are slowing down. Overseas, in Europe, the European Nanobiocom project is working on technology that can be used to repair bones that have deteriorated due to degenerative bone disease.[11] Millions of patients around the world are today suffering from degenerative bone disease, a disorder that is typically treated by implants and surgical grafts.

Researchers on the commission are seeking to develop a replacement for bone tissue that performs the same functions as healthy tissue. A balance must be struck, however. Seriously advanced stages of bone tissue degeneration require a substitute to completely duplicate both the structural and physiological properties of the decaying tissue tissue. But the technology must also be comfortable for the patient.

European researchers are developing "smart" materials that will assist in the repair and regeneration of cellular tissue in the bone. This new material acts on the genes in the cells. The material also adapts to the size and shape characteristics of healthy bones. Researchers are using nanoparticles to accomplish these medical goals.[11]

Elsewhere in Europe, researchers at Technalia are developing a biodegradable nanocompound for bone regeneration.[12] When a severe break in a bone occurs, bones take longer to regenerate. Thus, bone tissue engineers are working for the development of new material that induces bone regeneration and degrades as soon as the bone heals.

Ms. Beatriz Olade, research scholar at the Health Unit of Tecnalia, recently developed a new porous compound that interacts chemically and electrically with bone cells. The interaction with bone cells accelerates bone recovery.

Olade uses the polymer polylactic acid, a bioceramic hydroxyapatite, and carbon nanotubes for her work. The bioceramic furnishes calcium to the cells and thus stimulates their growth. Carbon nanotubes enhance mechanical properties of the polymer. Polymer polylactic acid presents biodegradable properties and is already used for medicinal purposes, and the polymer disappears as soon as the bone heals.[12]

Satisfactory results were found in both *in vitro* and *in vivo* experiments, and the results are encouraging and bone growth was observed after three weeks and, after sixteen weeks, bone showed mechanical and other properties of bone similar to healthy bone tissues.

Back in the U.S.A., the coolest innovation yet in nanotechnology may come from the “black ops” folks at the Pentagon.[13] The Defense Advanced Research Projects Agency (DARPA) has developed a biocompatible, nanobased compound for the treatment of fractures called “fracture putty.” Like the putty one uses around the house to repair broken objects, this putty is going to be used to repair broken bones.

Scientists from a number of research institutes collaborated with DARPA on the development of this new technology, including the University of Texas Health Science Center at Houston, Harvard University, and the University of Houston. Researchers will use peptide amphiphile (PA) for improving the bioactive features of the putty.

The Pentagon reports that tissue and bone injuries in American soldiers in Iraq and Afghanistan are all too common. In the field, these leg fractures often lead to amputation for young soldiers, fresh out of high school. The new technology, however, is leading treatment of fractures in the field and is a potential revolution in medicine, which researchers are calling *regenerative medicine*. The so-called fracture putty is targeted as a substitute for the damaged bone. The putty starts formation of natural bone and assists in healing the damaged surrounding tissues.[13]

References

[1] Nanotechnology May Help Grow Replacement Bone, FoxNews.com, May 10, 2006, http://www.foxnews.com/story/0,2933,194857,00.html.

[2] Nanotechnology Leads to Better Bone Implants, ZD Net, August 29, 2010, http://www.zdnet.com/blog/emergingtech/nanotechnology-leads-to-better-bone-implants/677.

[3] Those nano bones… are made for walking! University of Arkansas press release, http://dailyheadlines.uark.edu/Tian-Press-Release-July-2007.jpg.

[4] Murdoch University press releases, *http://media.murdoch.edu.au/tag/gerrard-poinern.*

[5] Orthovita Announces Presentation of Cortoss 24-Month Pivotal Study Results at the 25th Annual Meeting of the North American Spine Society, October 6, 2010, http://www.orthovita.com/investors/pressreleases.aspx.

[6] Orthovita Receives FDA Clearance for VITOMATRIX™ Bone Graft Substitute in Dental Applications, October 4, 2010, http://www.orthovita.com/investors/pressreleases.aspx.

[7] Orthovita Announces Publication of 24-Month Data from Cortoss Clinical Pilot Studies in Vertebroplasty and Kyphoplasty, September 20, 2010, http://www.orthovita.com/investors/pressreleases.aspx.

[8] Orthovita Notes Publication by The Lancet of Study Demonstrating Efficacy of Vertebroplasty vs. Conservative Treatment, August 11, 2010, http://www.orthovita.com/investors/pressreleases.aspx.

[9] Nano Bone Administered to Patients, R&D magazine, June 1, 2003, http://www.highbeam.com/doc/1G1–104611398.html.

[10] Competitive Technologies Licenses Nano Bone Biomaterial, http://www.competitivetech.net/.

[11] Nanotechnology Promises Bone Cell Regeneration, March 31, 2007, http://www.nanotechnologydevelopment.com/medical/nanotechnology-promises-bone-cell-regeneration.html.

[12] New biodegradable nanocompound for bone regeneration, August 3, 2010, http://www.nanotechnologydevelopment.com/research/new-biodegradable-nanocompound-for-bone-regeneration.html.

[13] Fracture Putty—A biocompatible nanobased compound to be developed for treating bone injuries, May 26, 2009, http://www.nanotechnologydevelopment.com/medical/fracture-putty-a-biocompatible-nanobased-compound-to-be-developed-for-treating-bone-injuries.html.

7

Chemical Substitutes

Nanotech Alchemy and Neowater

Neowater, the nanotechnology developed by Israel's Do-Coop Technologies, sounds like something out of one of Arnold Schwarzeneggar's Hollywood movies of the 1980s or 1990s. Something improbable, pure science fiction, something from the future, visiting us here, right now—nanotechnology-based water.

In interviews with the scientific press, however, Do-Coop CEO Eran Gabbai calmly and coolly positions Neowater as a logical extension of the chemical properties found within water itself.[1] But, as one of America's top science-fiction writers, Arthur C. Clarke, author of *2001: A Space Odyssey,* once reckoned, "any sufficiently advanced technology is indistinguishable from magic."

Neowater technology has acquired this aura by crossing chasms that have cost the pharma industry some $80 billion per year in dead-end R&D on chemical substitutes. Neowater solves a number of problems (i.e., inefficacy, high levels of toxicity, and lack of bioavailability) that harass new drug developers and have done so for years.

The nanotechnology chemical substitute—one of many emerging nanotech chemical substitutes being employed not only by medicine but by chemical engineers, too—really is derived from water. The simplicity of the innovation is its true beauty.

The human body transforms the water it consumes daily, and intracellular water demonstrates hydrophobic and hydrophilic compounds at the molecular level, handling them in a natural process. Gabbai says pharmaceutical researchers have never considered the possibility that the mysteries of natural water might hold the secret to the strangeness of intracellular water. Rather, pharma R&D has related to both kinds of water as barriers that block new pharma compounds, most of which are hydrophobic.[1]

Scientists have observed some well known properties of natural water for years without really understanding them. Intracellular water, which makes up 70 percent of our bodies, has decidedly different physical and chemical properties as compared to natural water.

The standard drug development process has included the use of a delivery vehicle other than water—usually dimethyl sulfoxide (DMSO), a liquid solvent. This reduces the necessary water to an inert state, as bulk or lab water; intracellular water is not even taken into the picture at all.

When drug testing transitions to the clinical phase, where the highly toxic DMSO must be replaced by some other delivery vehicle, even more problems arise. Whatever good results were achieved with the DMSO system were viewed as meaningless as the new vehicle carried the drug through *in vivo* testing. Then, quite often, the drug development process ends with efficacy failure or other unexpected result.

Pharma researchers keep changing the delivery vehicle in radical, new ways, and yet they expect solid data results, according Do-Coop scientists.

Oddly, conventional high-throughput screening does not account for the properties of the particular disease being targeted, and the intracellular water environment in which the disease is met. Drug companies commonly try to convert their entire IP library into a number of possible drug candidates for a match with a specific cellular receptor. The same tactic is tried repeatedly in the hopes that one will prove to be a hit drug.

This kind of testing happens millions of times for each study—an expensive and time-draining process. Despite the drain on R&D budgets, pharma leaders regard that kind of high-throughput screening as the industry standard.

The standard drug development process is based, however, on two false assumptions: (a) that a drug candidate and receptor should meet in an environment other than the cellular one in which the disease happens, and (b) that preclinical trials can rely on a vehicle hostile to cellular function in order to predict actual efficacy in *living* cells.

These false assumptions are based on one simple mistake, i.e., a failure to grasp the unique activity of water—natural water in general, and intracellular water in particular.

Gabbai and his team truly thought outside of the box when coming up with a solution—a challenge to the false premises of the drug development world. Rather than fighting water as the enemy in drug delivery, they wondered, "What if medicine were able to harness the unique properties of natural water to make it more like intracellular water? Would this new water become a drug delivery vehicle itself?"[1]

Do-Coop researchers transformed water from a barrier into an ally in drug development. The result is a liquid that mimics the intracellular water found in the human body, which they called "Neowater."

Physicist Richard Feynman first introduced the concept of nanotechnology to the scientific world, declaring that a successful technology must first and foremost relate honestly to known reality, because nature cannot be conned. The Neowater nanotech application embraces the Feynman Rule. Neowater shifts the physical and chemical properties of natural water, making it more like intracellular water, thus enabling R&D to deal more accurately with nature by allowing the use of a single vehicle for the entire drug development procedure.[1]

Neowater is compatible with the intracellular environment where a drug candidate and a receptor are destined to interact, and it removes the requirement for high concentrations of cosolvents and surfactants—other attempts to get past the earlier perceived problems with water, but, in the end, which cause a lack of predictability in drug efficacy testing.

Neowater is made of two parts hydrogen and part one oxygen—the same basic chemistry as conventional bulk water.

The primary difference is that water molecules in Neowater are collected in clusters, featuring different properties as compared to conventional bulk water. They key feature is the very large surface area of these clusters, which is achieved by introducing nanoparticles to the water. That is done with a proprietary process.

The post-production process of composing Neowater differs from regular water in that the CO_2 concentration is 10- to 100-fold higher. The enlarged surface area, due to the nanoparticles, also supplies structure during the liquid phase. This results in an enhanced ability of Neowater to disperse hydrophobic and hydrophilic compounds. What is more, it heightens bioavailability, efficacy, and stability, the company says.[1] The end product here is nontoxic, inert, and similar in characteristics to intracellular water. It is also remarkably stable.

The competitive threat—the primary one—is shrinking drug development pipelines. If companies don't come up with innovations, they will be out of business. That's where the business value of the Neowater innovation comes into play. Neowater, as a vehicle, facilitates the delivery of drugs. Since Neowater is a generic solution, it furnishes a delivery vehicle for many drug compounds with only minimal need for customization. This paves the way for new and extended intellectual property developments, based on expanded therapeutic possibilities.

Using Neowater as an aqueous vehicle, drug developers streamline their drug development process, moving from *in vitro* development to *in vivo* delivery. Then they can transition from early laboratory experiments to studies in humans, quickly, more cost effectively, and with fewer last-second surprises.

Gabbai says Do-Coop Technologies' production of Neowater is GMP compliant, and a Drug Master File (DMF) file was submitted to the FDA—DMF no. 20503. What is more, Do-Coop is currently in the process of having Neowater certified as *generally recognized as safe (GRAS)* by the FDA.

Neowater's DMF presents nontoxic studies that will enable pharmaceutical companies to utilize the technology with their drug products as part of the standard drug development and FDA registration processes. There is no need for a separate Neowater toxicity control, as a result of the DMF registration. This also allows developers to bring their products to market more quickly.

The use of Neowater in pharma applications and formulations also bolsters their therapeutic activity: better hydration and overall *in vivo* bioavailability, and better control in treatments requiring sustained-release formulas.

Gabbai reckons that Neowater provides the drug industry with a chemical substitute that can transform pharmaceutical and biotech research from a "game of chance" to a lucrative industry. Neowater can widen the possible therapeutic window, giving drug companies the chance to revive earlier drug development ideas that failed or were shelved because of efficacy and toxicity troubles. This recovers lost R&D investments from the past and can also unleash new therapeutic substances for the future, which will improve the quality of life for millions of patients.

Other Nanotech Solvents

Other nanotech-based solvents are coming into the development stream for use by medicine. Some background is needed to understand them. A carbon nanotube is a large molecule made of carbon atoms. The atoms are arranged in hexagons, much like chicken wire—the same arrangement as in graphite but rolled into a cylinder. Carbon nanotubes were discovered in 1991 by S. Iijima. These intriguing structures have sparked much excitement in the recent years.[2]

Currently, their physical properties are still being discovered and disputed. What makes this so difficult is that nanotubes have a very large range of electronic, thermal, and structural properties that change depending on the kind of nanotube.

Carbon nanotubes are skilled at entering the nuclei of cells and may one day deliver drugs and vaccines. Modified nanotubes so far have only been used to convey a small peptide into the nuclei of fibroblast cells. Researchers, however, are hopeful that the technique may one day form the basis for new anticancer treatments, gene therapies, and vaccines.

The 1996 Nobel Prize for Chemistry was won by Harold W. Kroto, Robert F. Curl, and Richard E. Smalley for their discovery in 1985 of a new allotrope of carbon, in which the atoms are ordered in closed shells. The atomic form was found to have the structure of a truncated icosahedron and was dubbed the *Buckminsterfullerene*, after the engineer Buckminster Fuller, who designed geodesic domes in the 1960s.

Back in 1990, physicists W. Krätschmer and D. R. Huffman produced isolable quantities of C60 by compelling an arc between two graphite rods to burn in a helium atmosphere and extracting the carbon condensate so creating an organic solvent.[2]

C60 are molecules approximately 1 nm in diameter, comprising 60 carbon atoms arranged as 20 hexagons and 12 pentagons: the configuration of a football. Hence, they find application as nanopharmaceuticals with a large drug payload in their cage-like structure. With development of various chemical substitutes for C60, it is now possible to create functionalized C60 with better drug-targeting properties.

C Sixty, a company based in Texas, has created a proprietary technology for scalable production of artificial membranes based on derived fullerenes.[2] The firm predicts that these artificial membranes, also known as *Buckysomes*, will prove useful in targeted delivery for a variety of different drugs to specific sites of disease in the body. C Sixty has a partnership with the University of Texas Health Science Center in Houston to optimize these drug delivery systems to deliver a new generation of anesthetics to critical body sites with greatly increased efficacy and safety. C60 for drug delivery is taking off and may even lead to a variety of applications of these interesting materials in the field of medicine.

Nanotech Alchemy: Gold and Silver

Something that has attracted a lot of attention in the scientific media lately has been a form of nanotech-based alchemy: inkjet printing of metal nanoparticles to create conductive metal patterns as an alternative to fabrication techniques like vapor deposition.[2] Research in this area concentrates on printing metal nanoparticle suspensions for metallization. Silver and gold nanoparticle suspensions are being ink-jet printed to build active microelectromechanical systems (MEMS) and flexible conductors for radio frequency identification (RFID) tags. Metals like silver and gold are the preferred nanoparticles for ink-jet formulations because they are good electrical conductors, and they do not cause oxidation problems later on that society must deal with as an after effect of manufacturing. But gold and silver still are too costly for most high-volume, ultra-low-cost applications such as RFID tags, which demand unit costs of less than one cent.

To print metals, one needs to prepare a metal ink. In terms of materials engineering, that means matching the properties of small metal particles with an ink-jet base fluid. What's more, for large-scale and low-cost industrial applications, the ink needs to be durable yet stable against air and humidity.

Although copper would be a decent choice to replace gold and silver, it, like other non-noble metal nanoparticles, tends to oxidize when put in contact with ambient air. But a new technique developed in Switzerland uses a flame spray synthesis in combination with a simple *in situ* functionalization step to synthesize graphene-coated copper nanoparticles that are air stable and easily handled. This research illustrates graphene's potential as a protective shell material for nanoparticles, enabling control and design of the chemical reactivity of non-noble metals.

"We have demonstrated that copper, as a low-cost, non-noble metal, can resist oxidation under ambient conditions if coated by graphene bi- or tri-layers," scientist Norman Lüchinger said.[3]

Lüchinger, a PhD at Wendelin Stark's Functional Materials Laboratory at ETH Zürich, in Switzerland, is first author of a recent paper, "Graphene-Stabilized Copper Nanoparticles as an Air-Stable Substitute for Silver and Gold in Low-Cost Ink-Jet Printable Electronics," published in *Nanotechnology*. The paper describes the suitability of graphene coating for the broader use of non-noble metal nanoparticles. The scientists in Zürich believe that carbon coatings offer an economically attractive route to the broader use of metal nanoparticles for ambient conditions.

Using copper as a substitute for silver or gold ink-jet printing demonstrates a first step in the development of commodity metal nanoparticle applications. Although currently inferior to the conductivity of printed gold or silver, the direct one-step synthesis of our carbon-coated copper particles and the use of well established ink chemistry allowed us to use the metal ink in off-the-shelf ink-jet printers. This makes metal printing available to virtually any laboratory without the need for special equipment.

Scientists point out that controlling the oxidation and corrosion of the technically most important non-noble metals through the controlled deposition of graphene coatings will enable a much broader use of common transition group metals in the form of nanomaterials.

Potential applications include RFID tags, which have to be ultra-low in cost for a broad application in industrial and consumer applications, the researchers said.

The copper metal dispersions synthesized are deep black, are stable for several weeks, and have a viscosity similar to classical ink-jet formulations. These can be directly filled in a commercial ink-jet printer's cartridge and processed using a conventional low-cost ink-jet printer.

One thing that still needs to be improved is the conductivity of the printed lines, which needs to be increased to become cost competitive with gold and silver. "The current non-optimized patterns have a relatively low, but sufficiently high, electrical conductivity to operate light-emitting diodes that were directly put onto the copper lines printed on a flexible polymer foil substrate," said Lüchinger. "Once we have sufficiently improved the conductivity, basically any application is imaginable."[3]

Nanotech Biocides

One area where nanotechnology is clearly making inroads as a substitute for hazardous chemicals is in the realm of coatings.[4] Coatings can create anti-adhesive surfaces that resist things (such as dirt) sticking to them or have biocidal properties to prevent living organisms from sticking to them like barnacles on the side of a ship. Biofouling/protective coatings for ship hulls help ship owners limit the growth of organisms of different kinds on surfaces in a marine or freshwater environment. Generally, coatings incorporating biocides—chemicals that kill organisms—are used. But scientists are now developing nanostructuring techniques for metal surfaces that prevent biofilm formation and bacterial adhesion as well as the attachment of larger organisms.[4] These technologies could also be deployed in hospitals, where they could prevent the spread of methicillin-resistant staphylococcus aureus (MRSA) and other opportunistic infectious agents, which are known to stick to surfaces in hospitals and cause nosocomial infections.

The news about these new surfaces appears in the European Parliament's Scientific Technology Options Assessment (STOA) committee's recently released a study on the role of nanotechnology in chemical substitution.[3] One of the other findings of STOA was that chemical "substitution" is not restricted to the replacement of a hazardous substance by less hazardous substances.

The report states that "due to the fact that nanotechnology is neither a group of substances nor a group of products, but an enabling technology, the way nanotechnology can provide solutions is more fundamental than just replacing the function of the substitute. It is assumed that nanotechnology provides new effects which are not based on chemical properties of the related material but on the physical properties caused by its size and shape. It can be used to develop completely different processes or different products which serve the same purpose but in a completely different way."

References

[1] Neowater: Not Science Fiction, Just Good Science. *Next Generational Pharmaceuticals* Issue 17, http://www.ngp-

harma.com/article/Neowater-Not-Science-Fiction-Just-Good-Science/.

[2] Elemental Carbon and Nanotechnology, http://www.nanopharmaceuticals.org/Bucky_Ball.html.

[3] Nanomercado, Low-Cost Nanotechnology Substitute for Gold and Silver in Printable Electronics. October 14, 2008, http://www.nanomercado.com/verNoticia/_Bi1IQ1fsmFE5Aa3VCIpHSbFtb73GYakdCaPv6cinPUqLAiiwOcOw0FVCwpYNG1qi.

[4] The Potential for Nanotechnology to Replace Hazardous Substances. *Nanowerk Spotlight,* July 12, 2007 http://www.nanowerk.com/spotlight/spotid=2212.php.

8

Nanotechnology, Hormones, and Hot Flashes

Hormone-based, transdermal therapies are one of the latest innovative therapies to emerge from the nanotechnology laboratory, according to Michael Moradi, an associate analyst at the market research firm, NanoMarkets. The technology is being used to treat hot flashes in menopausal women and for other gynecological therapies, and it has attracted significant attention from investment bankers who see it as a potentially huge market with the aging of the "baby boom" population.[1]

Indeed. Columbia, MD-based Novavax has developed two hormone replacement therapies, called Estrasorb®, which has already received Food and Drug Administration (FDA) approval, and Androsorb, which successfully completed Phase I human trials recently.

The number of these FDA-approved nanotech polymers available for use on skin is increasing rapidly, said Moradi.

The industry has been presented with opportunities to create new transdermal platform designs with enhanced "on-skin" properties and diffusion of active molecules compared to current patches. This is expected to result in smaller and less invasive skin patches, drug delivery devices that increase the universe of available drug candidates, according to Moradi's research. Electronics are even being integrated into patch-like platforms involving wound care, monitoring, and diagnostic methods.

NanoMarkets believes that the market for nano-enabled drug delivery systems will be quite large. Other experts agree. Merrill Lynch recently launched its Nanotech Index, which covers 22 companies that indicate in public documents that nanotechnology initiatives represent a significant component of their future business strategy.

Nelson M. Sims, president and CEO of Nanovax, noted that the investment bankers told him his firm was being included in the Nanotech Index because of its hormone therapy. "Our micellar nanoparticle (MNP) technology is one of Novavax's most exciting drug delivery platforms, and its commercial viability has already been validated with the recent FDA approval of Estrasorb, the company's proprietary topical emulsion for estrogen therapy," he said. "MNPs are proprietary oil and water nanoemulsions—less than 1 micron in diameter and smaller than a platelet or bacteria particle—that are designed to encapsulate and deliver alcohol-soluble drugs directly into the bloodstream when applied to the skin. Being composed of oil and water, MNP products have moisturizing and conditioning properties and are applied topically like a lotion. Other advantages seen with the delivery system are the ability for the active ingredient to enter the bloodstream quickly and deliver stable levels when applied daily."

Sims said the company is planning to debut other, proprietary hormone products that utilize the firm's MNP technology platform, and it is partnering with pharmaceutical companies to "introduce products that target these other significant markets."

Other company products include Nestabs, NovaNatal, and NovaStart, a line of prescription prenatal vitamins; Gynodiol, estradiol tablets, USP, an oral form of estrogen therapy; AVC Cream, a sulfanilamide vaginal cream for vaginal infections; and Analpram HC, a prescription corticosteroid and antipruritic product for hemorrhoids. The company's micellar nanoparticle technology involves the use of patented oil and water emulsions that are used as vehicles for the topical delivery of a wide variety of drugs and other therapeutic products, including hormones. In addition to Estrasorb, its recently approved topical emulsion for estrogen therapy, Novavax has several product candidates utilizing this technology in human clinical trials or in preclinical development, including Androsorb, a topical testosterone emulsion that has completed two Phase I clinical trials. Novavax has other drug delivery technologies, such as its Novasome and Sterisome technologies, that are being utilized to develop other products. Novasomes are also used as adjuvants to enhance vaccine effectiveness, and Sterisomes are used for the delivery of long-acting drugs in subcutaneous (under the skin) injections.[2]

The company is led by a cutting-edge scientific team.[3] Dr. Denis O'Donnell has been chairman of the board of directors of Novavax, Inc., since May 2000, and chief executive officer and director of Molecular Diagnostics, Inc., since February 2003. He was a general partner at Seaside Partners, LP, a private equity firm from 1997 to 2003, vice chairman of the board of directors of Novavax from 1999 to 2000, and senior advisor to Novavax from 1997 to 1998. Dr. O'Donnell was president of Novavax from 1995 to 1997 and vice president, business development of Novavax from 1992 to 1995. Currently, he is a director of Columbia Laboratories, Inc., and ELXSI Corp.

Dr. O'Donnell notes how the biopharmaceutical drug delivery company started off specializing in women's health care, and that choice has led to all of its innovations.

The company was spun off from an existing public company, an animal health care company called IGI, in December 1995. It has been a standalone company since then. There have been some other funding events, too, capitalizing the firm. Back in November 2003, the company raised $27 million in a takedown from a shelf registration. The funding was underwritten by C. E. Unterberg, Towbin.

The primary use of all the funds was to enable the firm to launch Estrasorb. Estrasorb was approved in October 2003 by the FDA.[3] The launch occurred in 2004.

As mentioned above, Estrasorb consists of a proprietary delivery technology called *micellar nanoparticles,* which is a platform delivery system developed by Novavax scientists. Micellar nanoparticles are used to entrap human estrogen, 17-

beta estradiol, and deliver it through the skin. The product resembles a lotion, and the FDA refers to it as an *emulsion*. The product is used to deliver estrogen through the skin to treat the acute symptoms of menopause in women, which are primarily hot flashes and night sweats.

The "estrogen-only" hormone replacement market in the U.S.A. today is approximately $1.5 billion in annual revenues. The market has been experiencing some increase in the last several years as a result of the Women's Health Initiative study through the National Institutes of Health (NIH). The NIH study concluded that estrogen replacement therapy should not be used for long-term treatment, but should be prescribed for the acute treatment of menopausal symptoms.

"There seems to be a lot of scientific, medical, very credible reasons for women using estrogen as the only treatment that will alleviate the symptoms of menopause, i.e., the hot flashes and the night sweats. The estrogen replacement market in the U.S.A. continues to be substantial. It will continue to remain large, as the average onset of menopause is age 51, and we're right in the midst of our baby boomers reaching age 51 and will continue to do so for a number of years going forward," said Dr. O'Donnell.[3]

The company has a marketing arrangement with King Pharmaceuticals and its women's health care group. Novavax has more than 65 dedicated OB/GYN reps who are currently selling Novavax products throughout the U.S.A.

They have existing relationships with the OB/GYN community, are experienced in selling women's health care pharmaceutical products, and are also teaming those 80 reps up with approximately 80 representatives from King Pharmaceuticals. The company therefore will approach the market with approximately 160 dedicated OB/GYN reps who will be detailing Estrasorb primarily to OB/GYN physicians, but also to internists, family practitioners, general practitioners, and clinicians who prescribe a fair amount of hormone and/or estrogen replacement therapy.

The company's drug development pipeline also has a lot of potential. The company has a testosterone product for women utilizing the same nanotechnology platform delivery technology. There are also a number of other products in the women's health care arena that are in early preclinical development. The company continues to be interested in establishing collaborations with other companies that may have products that could benefit from the use of our proprietary delivery systems.[3]

O'Donnell said there is a lot of interest these days in life science nanotechnology. "The simplest data point that I can give you is that Novavax is beyond the theoretical stage. We've actually developed a platform technology which has been patented, which has been through preclinical testing (Phase I, Phase II, and Phase III FDA human clinical trials), and has passed the greatest due diligence review in our industry, i.e., U.S. FDA drug approval," said O'Donnell. "So we are far ahead of a number of the companies that are either in the proof-of-principle stage or are early in development. We actually have something that's not only patented, but it is something that has gone through the rigors of FDA human safety and efficacy testing and has passed and been approved, and is now certified to be marketed to the U.S. population."[3]

The shareholder base of the company has been a noninstitutional base, primarily individual shareholders. There is a small institutional grouping of shareholders as well.

One leading analyst who covers Novavax is Ken Trbovich, formerly of CE Unterberg, Towbin.

In the long term, there are several points for those following Novavax to keep in mind. The company is addressing a large market in Estrasorb. The focus at Novavax is to take generic drugs, a drug like estrogen, 17-beta estradiol—a drug that the FDA is familiar with, including its pluses and minuses, its attributes, and its potential side effects—and transform it into a proprietary, patented, nanotechnology transdermal delivery system.

The patent on Estrasorb does not expire until 2015. When the FDA approved Estrasorb in 2003, it had an additional 12 years of patent protection on the first patent.

Nanotech Steroids in Sports

Hot flashes and other gynecological conditions are not the only focus of nanotech hormone therapy researchers. Back in 2008, at the Beverly Hills Hotel in California, a Growth Hormone Summit was held by the David Geffen School of Medicine at UCLA in conjunction with Major League Baseball (MLB) and the law firm of Foley and Lardner to examine the trend. Dr. Gary Green, professor of family medicine at the

UCLA medical school, chaired the conference of leading anti-doping experts and scholars. "Growth Hormone: Barriers to Implementation of HGH in Sports" addressed several scientific, legal, and ethical issues involving testing athletes for human growth hormone. Nanotech-based tests are coming to the market to test the urine of athletes thought to be "doping" with hormones to improve their on-field performance.

The big news at the Growth Hormone Summit was the increasingly viable urine test for human growth hormone that utilizes nanotechnology to identify urinary HGH markers. Don Catlin, CEO of Anti-Doping Research and professor emeritus at the UCLA School of Medicine, is collaborating with Lance Liotta, MD, PhD, of George Mason University to validate the utility of this test.

Don Catlin, a Los Angeles-based worldwide doping expert who managed blood testing for HGH at the Beijing Olympics, and Dr. Lance Liotta, a former pathology lab chief at the National Cancer Institute's Center for Cancer Research, recently launched a study to build upon Liotta's ability to identify isolated markers of HGH in human urine. "This is a ground-breaking step that'll change the game a bit," Catlin said.[4]

Since that time, Catlin has become increasingly confident that the new nanotechnology urine test for human growth hormone will offer the ideal solution for HGH testing in sports. Many athletes have objected to blood testing for HGH, which previously seemed to be the only anti-doping measure capable of detecting molecules of such small size. But the nanotechnology technique has apparently overcome that obstacle.[4]

"That is what is exciting about what Dr. Liotta is doing. He has a technique that we think will do that," Catlin said. "It is really brilliant."[4]

Ceres Nanosciences recently patented the nanotechnology method of collecting HGH in the urine for analysis by standard lab testing equipment. Virginia-based Ceres partnered with George Mason University in Fairfax, VA, and Italy's Istituto Superiore di Sanità when developing the technology, company CEO Thomas Dunlap said.[4]

Widespread adoption of the test probably depends on reviews by anti-doping authorities, leagues, and players unions. World Anti-Doping Agency representatives had a conference call with Ceres officials recently.[4]

But testing for HGH isn't the only nanotech innovation regarding human growth hormone. John Kressaty, the chemist behind "h" Serum, said that 3LAB Inc. recently invested £2.5 million to produce the nanotechnology necessary to allow HGH to penetrate the skin. "These hormones travel through the dermis via the hair follicles. About 80 percent of the growth hormone receptors in the skin are clustered at the base of the hair follicles," he says. "We view it as hormone-replacement therapy for the skin. Our clinical trials were done on sun-damaged, prematurely aged skin in Australia, and we found a 100 percent improvement in wrinkles and skin texture."[5]

Another expert, at the University of Reading, UK, is examining the use of nanotechnology to restore skin's youthfulness. Collagen growth has for a long time been seen as the ultimate cash prize for makers of anti-aging skin cream. Now there are clues as to how an ingredient in anti-wrinkle treatments may stimulate this growth and restore skin's elasticity.[6]

Ian Hamley is maneuvering to determine how the compound Matrixyl works by eyeing the nanoscale arrangement of its long carbon chain and the peptide of five amino acids attached to one end. Other compounds containing peptides made up of fewer amino acids tend to form cylindrical structures, with all the long chains pointing inward and the peptides outward. In Matrixyl, though, these cylinders are outnumbered by flat "nanotapes," in which the molecules are lined up in two layers with all the peptides on the upper and lower surfaces.[6] Surfaces formed by nanotapes may facilitate the buildup of collagen, he said. Hamley reckons this work will help research into regenerative medicine for injuries to collagen-containing tissue such as skin and the eye. Hamley's study was published in *Chemical Communications* (ANI). One caution, though: what happens when you actually apply it to skin is still unknown, observed Christopher Griffiths at the University of Manchester, UK.

A recent discovery that a vaginal gel consisting of nanoparticles carrying bee venom can serve both for contraception and for HIV prevention led to a grant from the Gates Foundation. Bee venom stimulates hormones such as cortisone and hydrocortisone.[7] This gel seems promising for improving sexual health for women. Condoms have been the number one choice to prevent the spread of HIV, but they depend heavily on male willingness to use them.

The concept came from Sam Wickline, a professor of medicine, cell biology, and physiology at the Washington University

School of Medicine. Wickline is one of 65 scientists to participate in the 2010 Grand Challenges Explorations Grant program, a $100 million program sponsored by the Bill & Melinda Gates Foundation.[7]

"Sperm and HIV are remarkably similar in their natural mechanism of genetic transmission," Wickline said in a news release. "Both need to fuse with their target cell in order to deliver their genetic payloads—DNA in the case of sperm, and RNA in the case of HIV." Although the gel is not as familiar to people as are condoms, doctors think that people can be informed about it easily.

There have been some recent trials of intravaginal microbicide gels that prevent against HIV and other STDs, and they've been found to be well accepted by women, so this is also something that probably could be implemented with great success.[7]

Research on the gel is in its early stages, but researchers believe the gel will be affordable. If researchers are able to prove that this works and that it can be manufactured on a large scale, the price would go down to affordable levels.[7]

Wickline said the gel contains a "trojan horse"—a lipid nanoparticle developed by Wickline and Gregory Lanza (a professor of medicine, biomedical engineering, and biology and biomedical sciences) that attracts sperm and HIV before destroying it using the bee toxin.

"The idea is to trick each [agent] to fuse with a synthetic Trojan horse—a nanoparticle that will overwhelm sperm and HIV in numbers and in destructive power," Wickline said.[7]

The bee toxin, known as *melittin,* comes from the honeybee *Apis mellifera*. The method has been shown to be safe for clinical use in human subjects.[7] "Cells readily take in melittin," Wickline said. "But once it gets in, it pokes holes in cell membranes to destroy the cells."

References

[1] Nanotechnology to Revolutionise Drug Delivery, in-PharmaTechnologist.com, March 7, 2005.

[2] Estrasorb, Novavax, Inc. (USA), October 2003, http://www.estrasorb.com/pdf/Esprit-ESTRASORB.pdf.

[3] *The Wall Street Transcript,* CEO Interview with Denis O'Donnell, Novavax, published March 29, 2004.

[4] *Steroids in Sports,* Nanotechnology HGH Urine Testing at 2008 Growth Hormone Summit, http://www.steroidreport.com/2008/11/12/nanotechnology-hgh-urine-testing-at-2008-growth-hormone-summit/.

[5] *The Times of London,* Nanotechnology Could Unlock Secret of Youthful Skin, http://www.timesonline.co.uk.

[6] Nanotechnology Could Unlock Secret of Youthful Skin, sify.com, Nov. 22, 2010.

[7] Vaginal gel with bee venom lowers transmission of HIV, http://www.nanotech-now.com/news.cgi?story_id=40870.

9

Immunosuppression and Nanotechnology

Clinicians regularly encounter patients who are affected by diabetes mellitus, infection, and wound healing issues, no matter what their specialty. The science of nanomedicine is starting to impact diabetes treatment and other immune-system-related disorders.[1]

Nanomedicine relies on components as small as 1/80,000th of the diameter of a human hair. At the scale of 1 nm—or 10 times the diameter of a hydrogen atom—materials and devices interact with cells and biological molecules in an interesting way.

Although the science of nanomedicine is still in its gestation, it has major potential applications in treating diabetes. For example, an implantable nanomedical device, which contains pancreatic beta cells from animals, could be used to treat the disease. This device provides a temporary restoration of the body's delicate glucose control feedback loop without the need of powerful immunosuppressants.

Another futuristic nanotechnology application is known as the SmartCell, which originated at the Massachusetts Institute of Technology (MIT). As glucose rises in the bloodstream, the structure of the SmartCell is eaten away. The breakdown of the SmartCell's protein matrix releases insulin.[1]

Other nanomedicine advancements and their potential in diabetes research and practice are emerging as well. These devices include the use of noninvasive glucose monitoring via implanted nanosensors. The key technologies include fluorescence resonance energy transfer (FRET) and fluorescence lifetime sensing as well as new nano-encapsulation technologies for sensors such as layer-by-layer (LBL) films. Scientists also observe that nano-encapsulation technology may achieve better insulin delivery in diabetes via improved islet encapsulation and oral insulin formulae.[1]

Additionally, nine clinical studies of nanomedicine-oriented applications of antiseptics, disinfectants, and antibacterial therapeutics recently were completed, looking at other areas of medicine related to immune system function. Six of the studies described synthetic nanomaterials with antibacterial activity that could moderate immune responses. The other three studies involved bio-inspired antibacterial nanomedicines, applications based on biological substances, which would have the same kind of moderating impact on the body's immune responses.

Most scientists believe nanotechnology will start seriously influencing medicine around the year 2020.[1] Nanotechnology has, however, already revolutionized many research areas. Orthopedic research is currently looking at the use of nanotechnology to address associated orthopedic implant design and tissue regeneration, and finding ways to ensure that the implants are not rejected is a key topic of research.

Orthopedic implants and tissue engineered constructs greatly depend on the biocompatibility of the material, so as to avoid immune system acceptance of the new tissues. Several techniques for patterning implant surfaces and efficiently constructing scaffolds for tissue engineering have emerged along these lines. These techniques include lithography, polymer demixing, phase separation chemical etching, electrospinning, and molecular self-assembly.[1]

Tissue engineering strategies have adopted composite scaffold approaches to mimic bone, because it is considered a natural composite of nanohydroxyapatite.[1] Nanoparticles including calcium triphosphate, bioactive glass, hydroxyapatite, synthetic chitin, chitosan, and biodegradable polymers have been molded into porous three-dimensional scaffolds for bone repair and regeneration purposes. This approach not only allows for mimicking bone in composition, it enables the incorporation of nanoceramics and enhances the material's mechanical strength and nanopographic features.[1]

Polymer ceramic matrices using carbon nanotubes offer high tensile strength, high flexibility, and low density, and they can be exploited to develop more successful orthopedic implant materials, which are accepted by the body's immune system. With the genesis of nanofabrication techniques, scientists may shape several biomaterials into nanostructures that simulate the native hierarchical structure of the bone. To be sure, all of these materials can trigger immune system responses in the body.

Nanoparticle-Decorated Immune Cells

Clinical trials using patients' own immune cells to target tumors have yielded promising results. However, this approach usually works only when patients also receive large doses of drugs designed to help immune cells multiply rapidly, and those drugs have life-threatening side effects. Now a team of MIT engineers has devised a way to deliver the necessary drugs by linking them to nanoparticles that are attached to the cells sent in to fight the tumor. Consequently, the immune-cell-stimulating drug reaches only its intended targets, greatly reducing the risk to the patient.

The new method may heighten the success rate of immune-cell therapies, which hold promise for treating many types of cancer, said researcher Dr. Darrell Irvine. Irvine led the team that recently published its results in the journal *Nature Medicine.* "What we're looking for is the extra nudge that could take immune-cell therapy from working in a subset of people to working in nearly all patients, and to take us closer to cures of disease rather than slowing progression," said Irvine. The novel approach may also be used to deliver other types of cancer drugs or to promote blood-cell maturation in bone-marrow transplant recipients, according to the researchers.[2]

To undertake immune-cell therapy, doctors take a type of immune cell called *T cells* from the patient, reengineer them to target the tumor, and then reinsert them into the patient. These T cells then hunt and destroy tumor cells. Clinical trials are ongoing in which immune-cell therapy is being tested in patients with ovarian and prostate cancers, as well as melanoma.

Immune-cell therapy is a very highly thought of approach to treating cancer. Major restrictions today include finding enough of the T cells that are specific to the cancer cell and then enabling those T cells to function properly in the patient. Researchers have attempted to inject patients with adjuvant drugs that stimulate T-cell growth and proliferation. These interleukins—naturally occurring chemicals that help promote T-cell growth—have produced promising results in human clinical trials, but interleukin therapy can induce severe side effects, including heart and lung failure, when the cells are placed into the blood stream in large doses.

Irvine and his colleagues took a new tack: to reduce toxic side effects, they transmogrified to lipid-based nanoparticles that they can link to sulfur-containing molecules normally found on the T-cell surface. The investigators loaded two interleukins—IL-15 and IL-21—into the nanoparticles and then put the nanoparticle–T-cell combo into mice with lung and bone marrow tumors. After the cells reached the diseased tissue, the nanoparticles degraded and released the drug over a seven-day period. Drug molecules attached themselves to receptors on the surface of the same cells that carried them, stimulating them to divide and grow.

Within two weeks, all of the tumors in the mice treated with T cells carrying the drugs disappeared. Test mice survived until the end of the 100-day experiment, while mice that received no treatment died within 25 days, and mice that received either T cells alone or T cells with injections of interleukins died within 75 days, researchers said.[2]

Irvine and his cohorts also showed they could connect nanoparticles to the surface of immature blood cells found in the bone marrow, which are commonly used to treat leukemia. Patients who receive bone-marrow transplants must have their own bone marrow destroyed with radiation or chemotherapy before the transplant, which leaves them vulnerable to infection for about six months while the new bone marrow produces blood cells. Drugs that accelerate blood-cell production along with the bone-marrow transplant may shorten the period of immunosuppression, thus making the process safer for patients, said Irvine. He reports successfully enhancing blood-cell maturation in mice.[2]

This study, which was supported in part by the National Cancer Institute, is detailed in a paper titled "Therapeutic Cell Engineering with Surface-Conjugated Synthetic Nanoparticles."

Backstory: Nanotech in Kidney Transplants

The granddaddy of immunosuppression drugs, sirolimus, was approved by the FDA in September 1999 for the prevention of transplant rejection. This appears to be the first nanotech drug to have received marketing approval in the U.S.A.[3]

The chemical sirolimus has many advantages over traditional calcineurin inhibitors, principally the fact that it has low toxicity in human kidneys. Patients maintained on calcineurin inhibitors tended to develop impaired kidneys and chronic renal failure. This has been avoided for years by prescribing sirolimus. However, on October 7, 2008, the FDA approved safety labeling revisions for sirolimus warning of the risk for decreased renal function associated with its use.[3]

Sirolimus is sometimes used in conjunction with calcineurin inhibitors and/or mycophenolate mofetil, providing a steroid-free immunosuppression regime. Nonetheless, impaired wound healing and thrombocytopenia is a possible side effect; thus, some transplant doctors prefer not to use it immediately after the transplant, administering it only after weeks or months have passed. The drug's optimal role in immunosuppression has not yet been determined and is the subject of a number of ongoing clinical trials, doctors said.[3]

A 2009 study, the lifespans of mice fed rapamycin (Wyeth Labs' brand name for sirolimus) were increased between 28 and 38 percent from the beginning of treatment, or 9 to 14 percent in increased lifespan. The treatment began in mice aged 20 months, akin roughly to 60 human years. Thus there is the possibility of an effective anti-aging treatment for humans at an already advanced age. Since the nanotech drug strongly suppresses the immune system, people taking rapamycin are more susceptible to dangerous infections, though. Doctors do not know whether rapamycin will have similar lifespan effects in man, and study authors caution the drug should not be used by the general population.[3]

Sirolimus can be used with coronary stents to prevent restenosis in coronary arteries following balloon angioplasty. The drug is formulated in a polymer coating that provides controlled release through the healing period following coronary intervention. Large clinical studies have demonstrated lower restenosis rates in patients treated with sirolimus-eluting stents as compared to bare metal stents, resulting in fewer procedures. A coronary stent is being marketed by Cordis, a division of Johnson & Johnson, under the trade name Cypher®. Some researchers think that such stents may increase the risk of vascular thrombosis.[3]

Sirolimus is currently being reviewed as a therapeutic option for autosomal dominant polycystic kidney disease (ADPKD). Reports indicate that sirolimus can reduce kidney volume and delay the loss of renal function in patients with ADPKD. There are other, off-label uses emerging for the drug, too.[3]

Sirolimus is also promising in treating tuberous sclerosis complex (TSC), a congenital disorder that leaves sufferers prone to tumor growth in the brain, heart, kidneys, skin, and other organs. Studies have linked mTOR inhibitors to remission in TSC tumors—subependymal giant-cell astrocytomas (SEGAs) in children and angiomyolipomas in adults. Many American doctors began prescribing sirolimus (Wyeth's rapamune) and everolimus (Novartis's RAD001) to TSC patients off-label. A number of clinical trials using both Rapamycin analogs, involving both children and adults with TSC, are under way in America.[3]

Thus far, the studies have shown that tumors often regrew when treatment stopped. Theories abound that the drug ameliorates TSC symptoms such as facial angiofibromas, ADHD, and autism, which are even further off label than the earlier mentioned uses.

The other effects of sirolimus may have a role in treating cancer. Researchers show that sirolimus restricts Kaposi's sarcoma in patients with renal transplants. Noted mTOR inhibitors like temsirolimus (CCI-779) or everolimus (RAD001) are being tested for use in cancers such as glioblastoma multiforme and mantle cell lymphoma, doctors said.[3]

A combination therapy of doxorubicin and sirolimus has been shown to inspire AKT-positive lymphomas into remission in rodents. Signaling promotes cell survival in Akt-positive lymphomas and moves to stop the cytotoxic effects of chemotherapy drugs like doxorubicin or cyclophosphamide. Sirolimus blocks Akt signalling and the cells lose their resistance to the chemotherapy. Bcl-2-positive lymphomas were completely resistant to the therapy; neither are eIF4E expressing lymphomas sensitive to sirolimus.

Rapamycin, like all immunosuppressive drugs, actually decreases the body's inherent anticancer activity. This allows cancers that would have been naturally destroyed to grow. Those on immunosuppressive medications have a 10- to 100-fold increased risk of cancer, as contrasted with the general population.[3] What is more, patients who currently have or have already been treated for cancer have a higher rate of tumor progression and recurrence than patients with an intact immune system.[3] These general considerations counsel prudence when exploring the potential of rapamycin to combat cancer, which is suggested by experiment. Rapamycin lowers the cancer risk in some transplant patients.[3]

Researchers recently observed a major improvement regarding retardation related to autism. The researchers showed that sirolimus regulates one of the same proteins that the TSC gene does, but in different parts of the soma. Doctors decided to treat mice three to six months old (adulthood in mice lifespans), and this increased the autistic mice's intellect to about that of normal mice in as little as three days.[3]

An experiment carried out on a rodent model of Alzheimer's indicates that rapamycin may have beneficial effects on learning and memory deficits seen in Alzheimer's.

Rapamycin is a macrocyclic polyketide isolated from *Streptomyces hygroscopicus* that has been shown to exhibit antifungal, antitumor, and immunosuppressant properties. The biosynthesis of the rapamycin core is accomplished by a type I polyketide synthase (PKS) in conjunction with a nonribosomal peptide synthetase (NRPS). The biosynthesis of the linear polyketide of rapamycin is organized into three multienzymes, RapA, RapB, and RapC, which contain a total of 14 modules. The three multienzymes are engineered in such a way that the first four modules of polyketide chain elongation are in RapA, the following six modules for continued elongation are in RapB, and the final four modules to complete the biosynthesis of the linear polyketide are in RapC. Then the linear polyketide is modified by the NRPS, RapP, which attaches L-pipecolate to the terminal end of the polyketide and then cyclizes the molecule yielding the unbound product, prerapamycin.

The central macrocycle, prerapamycin, is then changed by an additional five enzymes that lead to the final product, rapamycin. The core macrocycle is changed by RapI, SAM-dependent O-methyltransferase (MTase), which O-methylates at C39. After that, a carbonyl is installed at C9 by RapJ, a cytochrome

P-450 monooxygenases (P-450). After that, RapM, another MTase, O-methylates at C16. Lastly, RapN, another P-450 installs a hydroxyl at C27 immediately followed by O-methylation by Rap Q, a distinct MTase, at C27 to yield rapamycin.[3]

The genes responsible for rapamycin synthesis have been identified by doctors. As anticipated, three large open reading frames (OFRs) designated as rapA, rapB and rapC encode for three extremely large and complex multienzymes, RapA, RapB, and RapC. The gene rapL has been determined to be the source of code for a NAD+ dependent lysine cycloamidase, which converts L-lysine to L-pipecolic acid for incorporation at the end of the polyketide. A gene rapP is embedded between the PKS genes and translationally coupled to rapC encodes for an additional enzyme, an NPRS responsible for incorporating L-pipecolic acid, chain termination, and cyclization of prerapamycin. What is more, genes rapI, rapJ, rapM, rapN, rapO, and rapQ have been identified as coding for "tailoring" enzymes that modify the macrocyclic core. RapG and rapH have been identified to code for enzymes which have a positive regulatory role in the preparation of rapamycin through the control of rapamycin PKS gene expression.[3]

The 31-membered macrocycle begins as the loading domain is primed with the starter unit, 4,5-dihydroxocyclohex-1-ene-carboxylic acid, which is derived from the shikimate pathway. The cyclohexane ring of the starting unit is reduced during the transfer to module 1. A staring unit is then modified by a series of Claisen condensations with malonyl or methylmalonyl substrates, which are linked to an acyl carrier protein (ACP) and extend the polyketide by two carbons each. Each successive condensation, the growing polyketide is further modified according to enzymatic domains, which are present to reduce and dehydrate the polyketide thereby introducing the diversity of functionalities observed in rapamycin. After the linear polyketide is complete, L-pipecolic acid, which is synthesized by a lysine cycloamidase from an L-lysine, is added to the terminal end of the polyketide by an NRPS. The NSPS cyclizes the polyketide giving prerarpmycin, the first enzyme free results.[3] The macrocyclic core is then customized by a series of post-PKS enzymes through methylations by MTases and oxidations by P-450s to yield rapamycin, researchers said.

Rapamune/rapamycin costs about $37 a day for a 4-mg dose.

The U.S. patent for sirolimus has expired; thus it is available in generic formulations. Generic sirolimus is available from Canadian pharmacies, manufactured in India, for $20 a day, according to news reports.

References

[1] *Can Nanotechnology Have An Impact For Patients With Diabetes?* Volume: 22 Publication date: Nov. 1, 2009, Issue No. 11, Nov. 2009, Robert G. Smith, DPM, MSc, RPh, CPed, http://www.podiatrytoday.com/can-nanotechnology-have-an-impact-for-patients-with-diabetes.

[2] Therapeutic Cell Engineering with Surface-Conjugated Synthetic Nanoparticles. *Nature Medicine,* published at the National Cancer Institute's abstracts site, Sept. 20, 2010.

[3] Rapamune, an immunosuppressant indicated for the propylaxis of organ rejection in patients aged 13 years and older receiving renal transplants. Wyeth (USA), corporate web site, http://www.wyeth.com.

Index

About the Author

Eugene J. "Gene" Koprowski is an Emmy-award-nominated science, technology, and health journalist. He's written extensively for the last 25 years for FoxNews.com, *Forbes,* United Press International (UPI), *The Washington Post, The Wall Street Journal,* and other leading global media. He is a contributing editor to the *Encyclopedia of Health Services Research* (2009). Gene holds a master's degree from the University of Chicago and completed his undergraduate work at Northwestern University, Evanston, Ill. He earned a law degree with an emphasis on patent law from the Thomas Jefferson School of Law. Gene was appointed by Gov. Robert F. McDonnell to the Special Advisory Commission on Mandated Health Insurance Benefits in 2009, and his nomination was approved unanimously by the Virginia General Assembly for the four-year term in 2010.

CPSIA information can be obtained at www.ICGtesting.com
Printed in the USA
BVOW012233250612

293450BV00005B/6/P

9 781606 502488